Ethical Tensions from New Technology

The Case of Agricultural Biotechnology

_____

CABI Biotechnology Series

Biotechnology, in particular the use of transgenic organisms, has a wide range of applications including agriculture, forestry, food and health. There is evidence that it could make a major impact in producing plants and animals that are able to resist stresses and diseases, thereby increasing food security. There is also potential to produce pharmaceuticals in plants through biotechnology, and provide foods that are nutritionally enhanced. Genetically modified organisms can also be used in cleaning up pollution and contamination. However, the application of biotechnology has raised concerns about biosafety, and it is vital to ensure that genetically modified organisms do not pose new risks to the environment or health. To understand the full potential of biotechnology and the issues that relate to it, scientists need access to information that not only provides an overview of and background to the field, but also keeps them up to date with the latest research findings.

This series, which extends the scope of CABI's successful 'Biotechnology in Agriculture' series, addresses all topics relating to biotechnology including transgenic organisms, molecular analysis techniques, molecular pharming, *in vitro* culture, public opinion, economics, development and biosafety. Aimed at researchers, upper-level students and policy makers, titles in the series provide international coverage of topics related to biotechnology, including both a synthesis of facts and discussions of future research perspectives and possible solutions.

Titles available

1. **Animal Nutrition with Transgenic Plants**
   Edited by G. Flachowsky
2. **Plant-derived Pharmaceuticals: Principles and Applications for Developing Countries**
   Edited by K.L. Hefferon
3. **Transgenic Insects: Techniques and Applications**
   Edited by M.Q. Benedict
4. **Bt Resistance: Characterization and Strategies for GM Crops Producing *Bacillus thuringiensis* Toxins**
   Edited by Mario Soberón, Yulin Gao and Alejandra Bravo
5. **Plant Gene Silencing: Mechanisms and Applications**
   Edited by Tamas Dalmay
6. **Ethical Tensions from New Technology: The Case of Agricultural Biotechnology**
   Edited by Harvey S. James, Jr.

Ethical Tensions from New Technology

The Case of Agricultural Biotechnology

Edited by

Harvey S. James, Jr.

Professor of Applied Economics, University of Missouri

CABI is a trading name of CAB International

CABI
Nosworthy Way
Wallingford
Oxfordshire OX10 8DE
UK

CABI
745 Atlantic Avenue
8th Floor
Boston, MA 02111
USA

Tel: +44 (0)1491 832111
Fax: +44 (0)1491 833508
E-mail: info@cabi.org
Website: www.cabi.org

Tel: +1 (617)682-9015
E-mail: cabi-nao@cabi.org

A catalogue record for this book is available from the British Library, London, UK.

Library of Congress Cataloging-in-Publication Data

Names: James, Harvey S. (Harvey Stanley) Jr., editor.
Title: Ethical tensions from new technology : the case of agricultural
    biotechnology / edited by Harvey S. James, Jr.
Description: Boston, MA : CABI, 2018. | Series: CABI biotechnology series ; 6
    | Includes bibliographical references and index.
Identifiers: LCCN 2018016831 (print) | LCCN 2018026246 (ebook) | ISBN
    9781786394651 (ePDF) | ISBN 9781786394668 (ePub) | ISBN
    9781786394644 (hbk : alk. paper)
Subjects: LCSH: Agricultural biotechnology--Moral and ethical aspects.
Classification: LCC S494.5.B563 (ebook) | LCC S494.5.B563 E84 2018 (print) | DDC 338.1/6--dc23
LC record available at https://lccn.loc.gov/2018016831

ISBN-13: 978 1 78639 464 4 (hbk)
        978 1 78639 465 1 (PDF)
        978 1 78639 466 8 (ePub)

Commissioning editor: David Hemming
Editorial assistant: Emma McCann
Production editor: Tim Kapp

Typeset by SPi, Pondicherry, India
Printed and bound in the UK by CPI Group (UK) Ltd, Croydon, CR0 4YY

Contents

Contributors

Philipp Aerni is Director of the Center for Corporate Responsibility and Sustainability (CCRS) at the University of Zurich. He completed his PhD at the Institute of Agricultural Economics, ETH Zurich and subsequently continued research at Harvard University, ETH Zurich, the University of Bern and the Food and Agriculture Organization (FAO) of the United Nations. His research interests focus on issues related to the ethics and politics of sustainable agriculture. E-mail: philipp.aerni@uzh.ch

Rachel A. Ankeny serves as Associate Dean Research and Deputy Dean in the Faculty of Arts at the University of Adelaide, where she also convenes the Food Values Research Group and the new Public Engagement in Science and Technology Adelaide (PESTA) Research Cluster. Her research explores a range of fields, including food studies, bioethics, history and philosophy of the biological and biomedical sciences, public engagement in science, and migration studies. She has received several competitive grants for research on these topics. She is editor-in-chief of *Studies in the History and Philosophy of the Biological and Biomedical Sciences*. E-mail: rachel.ankeny@adelaide.edu.au

Bartosz Bartkowski is research fellow in the Department of Economics at the Helmholtz Centre for Environmental Research in Leipzig, Germany. He received his PhD in economics from Martin Luther University Halle-Wittenberg, Germany. His research has focused on economic valuation of biodiversity and ecosystem services, particularly deliberative monetary valuation; governance of agricultural soils and novel agricultural technologies; theory and operationalization of sustainability; degrowth and limits to growth; environmental ethics and deliberative democracy. He is a member of the European Society of Ecological Economics. E-mail: bartosz.bartkowski@ufz.de

Heather J. Bray is an interdisciplinary scholar in the School of Humanities at the University of Adelaide. Her research explores how attitudes to science in agriculture and food are shaped socially, culturally and historically. Her areas of focus are GM crops and food, and livestock production. She has recently returned to full-time research after working for over 10 years in science communication, developing a community engagement program for agricultural research centers that use complex and controversial technologies, including gene technology. Her background is in agricultural science and she has worked as an animal scientist in both Australia and the Netherlands. E-mail: heather.bray@adelaide.edu.au

Harvey S. James, Jr. is Professor of Applied Economics and Associate Division Director in the Division of Applied Social Sciences at the University of Missouri. He is also editor-in-chief of the journal *Agricultural and Human Values*. He teaches courses in microeconomics, new institutional economics and applied ethics. His research focuses on applied ethics and the economic foundations of trust, ethics and happiness. He has a particular interest in the organizational structure of firms and markets and ethical issues relating to contracting and market competition. Much of his research is directed at the agrifood sector. E-mail: hjames@missouri.edu

Bradley M. Jones is a PhD student in Cultural Anthropology at Washington University in St. Louis. His research explores alternative agriculture, environmental social movements, solidarity economies, and neo-agrarianism in the USA, with a particular interest in young and beginning farmers. His research, reviews, and encyclopedia entries have appeared in *Food, Culture, and Society*; *CuiZine: the Journal of Canadian Food Culture*; *Digest: A Journal of Foodways and Culture*; *Culture and Agriculture*; *Gastronomica*; and the *Encyclopedia of Food and Agricultural Ethics*, among others. Brad is the founding editor of the *Graduate Journal of Food Studies* and the current President of the Graduate Association for Food Studies. E-mail: bradleyjones@wustl.edu

Deepthi E. Kolady is an Assistant Professor in the Department of Economics at South Dakota State University (SDSU). Her research interests span the fields of agricultural economics, natural resource economics, and development economics. She has a special interest in understanding how technology solutions, market access, and public policy affect people's decision making and contribute to economic development. Prior to joining SDSU, she worked at the World Bank, International Food Policy Research Institute, and Cornell University. E-mail: deepthi.kolady@sdstate.edu

Jane Kolodinsky is Professor and Chair of the Department of Community Development and Applied Economics and Director of the Center for Rural Studies at the University of Vermont. Her interests focus primarily on consumer affairs but have spanned multiple arenas, including community entrepreneurship, international development, affordable housing, rural affairs, food systems and health care. E-mail: jane.kolodinsky@uvm.edu

Katie M. MacDonald is an adjunct professor at the University of Guelph. Her work draws on the sociology of food and eating and political economy to bring a critical lens to modern North American food and agricultural systems. Her research interests include food environments, value chains, intensive animal agriculture, animal welfare, sustainability of natural resources, animal disease and ethics of care, and more recently, the perception and viability of lab-grown and faux 'meats'. E-mail: kmacdo08@uoguelph.ca

Kelly A. McKinley is a PhD student in the Department of History and a member of the Food Values Research Group at the University of Adelaide. Her current research examines the nature and impact of conflict and social movement activity around genetically modified crops and food in Australia, which is part of an ARC-funded Discovery Project documenting the history of GM in Australia. She is a 2018 recipient of the Australian Academy of Science's Moran Award for History of Science Research. E-mail: kelly.mckinley@adelaide.edu.au

Desmond Ng is an Associate Professor of Agribusiness and Strategy Management at Texas A&M University. He teaches undergraduate and graduate courses in managerial economics and strategic management. His research interests include areas involving: innovation processes, entrepreneurship, behavioral psychology, ethics and stakeholder management. E-mail: dng@tamu.edu

Frauke Pirscher is Senior Research Fellow at the Department of Agricultural, Environmental and Food Policy at the Martin Luther University-Halle Wittenberg. She received her PhD in Agricultural Economics from the University of Hohenheim, Germany. Her research focuses on agricultural and environmental ethics and environmental economics. E-mail: frauke.pirscher@landw.uni-halle.de

Roberto Quiroz recently retired as a project director in the Crop and Systems Sciences Division at the International Potato Center in Lima, Peru. He is a researcher with formal education in basic and applied sciences and has over 30 years of field experience in international research and development institutions as well as more than 14 years in higher education, putting science at the service of people. He is experienced in conducting and leading basic and translational research, focusing on understanding the interaction between climate and agriculture and how farmers can cope and adapt to extreme events in the Americas, Asia and Africa, and in using precision agriculture tools and methods to benefit small-holders. E-mail: raquirozguerra@gmail.com

Dane Scott is Director of Mansfield Program in Ethics and Public Affairs and Associate Professor of Ethics in the Department of Society and Conservation in the Franke College of Forestry and Conservation, University of Montana. He received his PhD in philosophy from Vanderbilt University and has a BS in soil science from the University of California-Riverside. His research focuses on conservation ethics and ethical issues arising from the use of emerging technologies to address environmental problems. He has been the principal investigator on two successfully completed National Science Foundation projects and has written numerous articles on ethics and agricultural biotechnology and climate engineering. E-mail: dane.scott@mso.umt.edu

Shivendra Kumar Srivastava is a researcher at ICAR-National Institute of Agricultural Economics and Policy Research, New Delhi, India, and is presently deputed to NITI Aayog. His research interests include sustainable agriculture production, food security, and policy analysis. Previously he has been a visiting scholar at Cornell University and the Australian Bureau of Agricultural and Resource Economics and Sciences. E-mail: sk.srivastava@icar.gov.in

Debra M. Strauss is Professor of Business Law at the Charles F. Dolan School of Business, Fairfield University. She received her law degree at Yale Law School. She is a Food and Drug Law Institute Scholar, who teaches courses in the legal environment of business, international law, and business law and ethics. In addition, she collaboratively developed and taught for the Honors Program, 'Interdisciplinary Inquiry: Critical Environmental Policy and Legal Challenges for the 21st Century'. E-mail: dstrauss@fairfield.edu

Insa Theesfeld is Professor in Agricultural, Environmental and Food Policy at the Martin Luther University Halle-Wittenberg in Germany. She is an agricultural and institutional economist who developed an interest in governance questions of various shared natural resources. A significant strand of her work explored the linkages between one or more natural resources, emerging from the prevailing property rights structure. She is particularly concerned about the compatibility between formal institutions and society's norms and values that influence effective policy implementation. She is a known 'Commons' scholar, serving in the council of the International Association for the Study of the Commons since 2012. E-mail: insa.theesfeld@landw.uni-halle.de

Johannes Timaeus is a biologist by training. His main professional interests are crop plants, crop plant diversity and the diverse relationships between crop plants and human culture. He is currently conducting a PhD project about diversifying agroecological systems by species mixtures at the Department of Ecological Plant Protection at Kassel University. He is also engaged in hands-on community gardening projects and the German Association for the Conservation of Crop Plant Diversity (Verein zur Erhaltung der Nutzpflanzenvielfalt). E-mail: johannes.timaeus@uni-kassel.de

Corinne Valdivia is Professor of Agricultural and Applied Economics in the Division of Applied Social Sciences at the University of Missouri. She teaches graduate and under-graduate courses in International Agricultural Development and Policy and oversees the Graduate Interdisciplinary International Development Minor. Her research focuses on how people and communities adapt to transformational changes brought about by innovations, climate change, and migration, using interdisciplinary, collaborative, and translational research processes in East Africa, the Andean region of Latin America, and rural communities of the Midwest. E-mail: valdiviac@missouri.edu

Duane Windsor is Lynette S. Autrey Professor of Management in Rice University's Jones Graduate School of Business. He edited *Business & Society* between 2007 and 2014. He has also published in *Business Ethics Quarterly* and *Journal of Business Ethics*. He has served as elected program chair and head of the International Association for Business and Society (IABS) and of the Social Issues in Management Division of the Academy of Management. He is an associate editor on the second edition of Sage's *Encyclopedia of Business Ethics and Society*. His research focuses on corporate social responsibility and stakeholder theory. E-mail: odw@rice.edu

Introduction
Ethical Tensions and New Technology: An Overview in the Context of Agricultural Biotechnology

Harvey S. James, Jr.

Division of Applied Social Sciences, University of Missouri, Columbia, Missouri, USA

A distinguishing characteristic of the human species is our ingenuity. We create, we learn, we invent and we evolve. We advance science and construct technologies derived from that science. Sometimes the reverse occurs – our building, experimenting, tinkering and creating help us conceive of new scientific possibilities. The drive to invent and develop new technologies is relentless. Sometimes we do this because we can. Sometimes we do this in response to real or perceived problems. Sometimes we do this out of a genuine desire to improve the wellbeing of others or the world. Sometimes we do this for fame or profit. Regardless of the reasons or motivations, the use and development of technologies can be applauded as well as condemned, and so the introduction of new technologies is often controversial. If we can do something, such as push technological frontiers outward, then should we? And if we want to push those frontiers outward, then what do we gain and give up in doing so?

New technologies, and new uses of existing technologies, produce ethical tensions. An ethical tension is created when the technology creates a conflict of interests, values or rights. Interests are things we care about. Values are ends in themselves or ideals to which we aspire. Rights are entitlements to things we obtain or do. New technologies create ethical tensions because they have differential effects on the interests, values and rights of others. Ethical tensions occur between individuals or groups and hence result in winners and losers, but they also exist within individuals. For example, an inventor might face a conflict between her interest in commercializing an invention and her interest in seeing that its benefits are distributed as widely as possible.

This book is an exploration into ethical tensions that new technology creates, with a particular focus on agricultural biotechnology and genetically modified (GM) food. Ethical tensions resulting from or related to the genetic modification of crops and plants and their supportive gene technologies have been apparent for decades in popular and scholarly publications, media reports and social network commentaries.

The term 'agricultural biotechnology' is frequently used synonymously with 'genetic modification' and 'genetic engineering', even though technically these are different things. 'Biotechnology is a collection of technologies that capitalize on the attributes of cells, such as their manufacturing capabilities, and put biological molecules, such as DNA and proteins, to work for us' (BIO, 2006, p. 1). By this definition, using yeast to make bread constitutes a biotechnological process. In contrast, genetic modification or genetic engineering refers to the manipulation of an

E-mail: hjames@missouri.edu

organism's DNA in order to alter characteristics of the organism. Selective breeding, which humans have been doing with plants and animals for millennia, falls into this category. The controversy about biotechnology stems from attempts to directly manipulate DNA in a laboratory and because genetic material and traits from one species are being introduced into another species. Agricultural biotechnology is the application of genetic modification techniques to crop plants and animals used directly or indirectly in the production of food and feed for humans and animals.

Why focus on agricultural biotechnology in order to examine ethical tensions from new technologies? Haven't we already addressed the major problems and issues relating to agricultural biotechnology? After all, agricultural biotechnology is not new, and many books have already been written about ethical issues arising from the development and use of agricultural biotechnology (see, for instance, Holland and Johnson, 1998; Ong and Collier, 2005; David and Thompson, 2008; Twine, 2010; Qaim, 2016). There are several reasons why a renewed focus on agricultural biotechnology can provide insights into ethical tensions created by new technology.

First, even though agricultural biotechnology has been around for a while, there are persistent controversies about its development and use. For example, many scientists argue (some might say persuasively) that current GM foods are safe to eat. The report by the United States National Academies of Sciences, Engineering and Medicine published in 2016 on genetically modified crops is perhaps the most definitive contemporary assessment of the safety of GM technology when applied to food crops. While acknowledging 'that there are limits to what can be known about the health effects of any food, whether non-GE or GE' (NAS, 2016, p. 17), the report concludes

> the research that has been conducted in studies with animals and on chemical composition of GE foods reveals no differences that would implicate a higher risk to human health from eating GE foods than from eating their non-GE counterparts.

Long-term epidemiological studies have not directly addressed GE food consumption, but available time-series epidemiological data do not show any disease or chronic conditions in populations that correlate with consumption of GE foods. The committee could not find persuasive evidence of adverse health effects directly attributable to consumption of GE foods. (p. 236)

But critics of agricultural biotechnology are not singularly concerned about food and crop safety. Intellectual property rights, the adequacy of regulations, labeling and corporate power over the development and use of GM technologies are examples of important, relevant and unresolved problems. Proponents of agricultural biotechnology who push safety claims when opponents argue about inadequate regulation or corporate power miss this point. Furthermore, one of the main arguments proponents of biotechnology make is that it is needed to feed the world (see, for instance, Daniels, 2017). This argument might be relevant in the case of some developing countries, but it is not necessarily relevant in developed countries where food availability is less of a problem than other food security concerns, such as access and adequacy. Indeed, the concerns raised by many critics of agricultural biotechnology are about GM use in developed countries, such as the USA and European Union. Some of these concerns have their counterparts in other new or emerging technologies, such as unmanned aerial vehicles and self-driving cars.

Second, agricultural biotechnology is being used to develop new agricultural products and production processes that continue to create challenges. The 'first generation' of plant biotechnology focused on herbicide tolerance and insect resistance, which lowers costs of production and increases yields and returns for farmers. In this case, the harvested GM crops were not designed to be different from their non-GM counterparts. More recently, 'second generation' GM techniques are being used to produce enhanced nutritional qualities or improved processing characteristics of crops. For example, genetic engineering has been used to increase the presence of pro-vitamin A (beta-carotene) in rice (Ye *et al.*, 2000), resulting in a product known as 'Golden Rice' (see Schaub *et al.*, 2005). According to the World Health Organization (2016, n.p.), other crop biofortification projects include 'iron-biofortification of rice, beans, sweet potato, cassava and legumes; zinc-biofortification of wheat, rice, beans, sweet potato and maize; provitamin A carotenoid-biofortification of sweet potato, maize and cassava; and amino acid and protein-biofortification of sorghum and cassava'. By biofortifying foods that people are accustomed to eating, proponents argue that there is greater opportunity to improve the health and wellbeing of adults and children suffering from chronic malnutrition around the world, especially in 'rural areas, which are not as easily reached by conventional programs' (Stein, 2015, p. 165). The 'third generation' of GM crops 'act as "factories", producing industrial goods, pharmaceuticals, and other products more efficiently and cheaper than traditional approaches' (Stewart and Knight, 2005, p. 522). For example, with third-generation agricultural biotechnology, children will be able to eat fruits and vegetables that contain 'vaccines and antibodies for a wide range of diseases like rabies, traveler's diarrhea, cholera, hepatitis B, antibodies to fight cancer, and tooth decay, and therapeutic proteins for cystic fibrosis, liver disease, and hemorrhage' (Stewart and Knight, 2005, p. 522).

The second- and third-generation developments of agricultural biotechnology seem to amplify ethical tensions that were identified with previous GM developments in agriculture. For example, since the genetic modification of crops is designed to alter substantially the resulting crop or animal product, there are renewed and additional concerns about the long-term health, safety and effectiveness of biofortified food crops or crops with pharmaceutical properties. In other words, even if first-generation GM crops are determined to be safe to consume, that does not mean that new and enhanced crops will not have adverse health effects on consumers or result in other unanticipated problems. Others argue that these second- and third-generation developments do not solve problems of malnutrition or chronic diseases but rather distract us from necessary reform efforts that focus on the root problems, such as poverty. For instance, Scott (2011) explains how Green Revolution technologies and Golden Rice today can be subjected to the 'technological fix criticism'. This criticism arises from the claim that complex social and environmental problems targeted for technological solutions are inappropriately defined as technological in nature and that the introduction of new technologies will not solve the problem and might in fact create new, unintended consequences. As an illustration, Scott explains how the Green Revolution introduced unforeseen negative social and environmental consequences and suggests that Golden Rice 'is a textbook example of a technological fix' that 'avoid[s] the problem of changing people's attitudes and behavior' (pp. 222, 223). In other words, Golden Rice masks the symptoms of malnutrition and does not solve the underlying problem causing it in developing countries.

Third, new technologies are being developed that relate to or derive from biotechnology and that have applications in food and agriculture, but that raise a host of intriguing ethical challenges. In other words, what we think we know about 'agricultural biotechnology' today might not fully prepare us for the scientific and technological innovations, and their associated ethical tensions, that are yet to come. For example, consider Clustered Regularly Interspaced Short Palindromic Repeats, or CRISPR, which is a technique of selecting and editing specific stretches of genetic code with greater precision than previous genetic modification techniques. Among other applications, Ledford (2015) reports that CRISPR has been used 'to fix a disease-causing mutation in an adult animal – and an important step towards using the technology for gene therapy in humans' (p. 21) and 'to engineer petite pigs and to make disease-resistant wheat and rice' (p. 22).

CRISPR is controversial for several reasons. The science behind the technology 'is moving much faster than the regulatory mechanisms that govern it' (Jorgensen, 2016, n.p.). The technology is easier and less costly to use than more traditional gene editing techniques, thus raising the prospect that more labs than can be regulated will use it to manipulate genomes of living organisms, including those of humans. Indeed, some researchers have already begun using the technology to modify human embryos (Cyranoski and Reardon, 2015). If more laboratories are using the technology because of the perception (or reality) that it is 'cheap' and 'easy', then there is greater risk that inexperience or negligence can result in mistakes in the design of the gene targeting system that will have unintended negative effects on living organisms or larger ecosystems. Additionally, because of the precision in editing DNA sequences using CRISPR, it becomes 'more difficult for regulators and farmers to identify a modified organism once it has been released', which in turn raises problems with identity preservation in farm crops and labeling requirements for genetically modified foods (Ledford, 2015, p. 22). There is also a major patent battle over conflicting patent claims to various versions of CRISPR technology, suggesting that ethical issues relating to intellectual property rights will persist.

The Four Arenas within Which Ethical Tensions Occur

A focus on ethical tensions encourages us to consider not only obvious ethical challenges, such as conflicts of financial interests and the potential for added risks or harm from new technology, but also fault lines and pressure points that can become problematic under the right conditions. There are four main arenas where new technologies generally, and agricultural biotechnology in particular, create ethical tensions, fault lines and pressure points. Each chapter in this edited volume focuses on at least one of these arenas (and many chapters address overlapping arenas).

The first arena is in the domain of public opinion and public interests. New technologies create ethical tensions among the interests, values and rights of technology innovators and those of the public, since what is good for the innovator might not be good for the general public. Simply stated, are

science and technological developments in agriculture proceeding according to public opinion and in the public interest or are they the result of what is in the best interest of the agrifood industry? How do we balance the competing interests of public and private interests? Should universities or companies restrain their scientific pursuits and technological developments until the public has weighed in on the matter, or should these proceed regardless of what public opinion is? What is the current state of public opinion about agricultural biotechnology, and does it matter? This last question is particularly difficult to answer, because there are so many varied interests and perspectives among people and across societies. Assessing and measuring public opinion therefore raises its own set of ethical challenges. A related issue is what gets communicated to the public and how that communication occurs. Developers of technology and public commentators can have a significant impact on public opinion by how they describe or frame technologies and their benefits and risks, and so communication raises many ethical tensions associated with new technologies.

The second arena focuses on policy and regulation, because these have direct effects on the interests, values and rights of stakeholders. New technologies raise questions about the adequacy of existing policies and regulations. Are existing regulations adequate for the kinds of technologies we are developing? When governing rules have to change, how should those changes be determined and implemented? Who wins and who loses when regulations change or do not change? The problem here is not just about the adequacy of existing regulations but also includes a consideration of the interests and values of governing bodies and policy makers. Should they reflect the concerns and interests of the general public or more narrow interests relating to the new technologies? Furthermore, there are differences in how agricultural technologies are regulated across countries. Is there a convergence in policy approaches to agricultural biotechnology internationally, and should there be? Is policy directing technological development where it is needed? How do we know? While regulations can and often are motivated by ethical considerations, often they are not. Winners and losers, interest groups and stakeholders vie for regulatory favors. Understanding better the ethical tensions relating to agricultural biotechnology provides insights into how regulatory battles might unfold for other technologies.

The third arena encompasses the relationship between new technologies and the social, economic or environmental problems they are often designed to solve. For example, ethical tensions arise from the way complex problems are defined and how technologies are utilized to resolve them, because the framing of problems and their solutions can create conflicts of interests, values or rights. Are new technologies solving problems or are they creating new ones? Is the introduction of a new technology the best approach to solving a problem, or do attempts to introduce new technologies direct our attention away from more effective or even necessary solutions? Some commentators identify this concern as the 'technological fix criticism', as explained above. The technological fix criticism is not a trivial issue. We face highly complex problems, some of which can be linked directly to the use of technology. Thus, introducing newer or more advanced technologies to solve problems caused by older technologies begs the question of whether we are focusing on the right types of solutions. Is it prudent to implement a solution of the same type that caused the problem to be solved? Consider this question: What agricultural technologies are widely acknowledged as being successes that have not introduced new problems? Related to the technological fix issue is the question of how technology relates to users and affects how users see and interact with reality. For example, our perception of the world differs depending on whether we experience an ocean beach with our senses, such as eyes, ears and nose, or whether we watch an ocean scene on a television. Developers of technology therefore have a significant influence on how users experience and understand reality, and this in turn creates the potential for serious ethical tensions and challenges.

The fourth arena concerns the interaction between new and old technologies. Sometimes new technologies complement older technologies (e.g. internet access obtained through existing telephone lines). More often, new technologies replace their older counterparts (e.g. mobile cell phones replacing landline phones) in a process that the Austrian economist Joseph Schumpeter described as 'creative destruction'. In this arena, ethical tensions derive from the interests, values and rights of those who develop and use existing technologies and those who develop and potentially benefit from new technologies. Should technologies be developed that replace existing technologies, especially when what exists seems to work well? Are there times we should restrain new technology development in order to protect the interests of some stakeholders? Should those developing new technologies take into consideration the interests and concerns of existing technology producers and users? When is technology developing too quickly?

These arenas can overlap. For example, a focus on the public interest relates to both existing and new technologies, and it affects how social, economic and environmental problems are defined. Public opinion and interest also interact with policy making. Similarly, the technological fix criticism arises because problems defined as technological often leave out social, cultural and other relevant contexts important to the public interest.

The questions raised here are important. Not all of them are covered in this book. But contributing authors in this volume address many of these issues and provide insightful commentary about ethical tensions arising from agricultural biotechnology.

It is important to note that none of the contributors in this volume is against biotechnology or genetic modification or, more generally, new technology. This is not an anti-biotechnology book. Indeed, all scholars contributing here acknowledge the benefits that agricultural biotechnology provides and how it can be used to improve food and the food system and advance the interests of humans, animals and the environment. But a favorable opinion does not mean one is blind to ethical challenges, and raising ethical concerns does not make one an opponent.

There are real challenges that new technologies create and that exist with the use of agricultural biotechnology. Honest inquiry requires that we recognize and give consideration to ethical tensions arising from the introduction of new technology, and that is what this book does.

Contributions in this Volume

The book is divided into five parts; the first four parts correspond to each of the four arenas described above.

The first section contains four chapters focusing on ethical tensions relating to public opinion and interest. In the first chapter, entitled 'Ethical Tensions from a "Science Alone" Approach in Communicating Genetic Engineering Science to Consumers', Kolodinsky considers 'the ethical tension between the need to communicate the risks and benefits of science and technology to the public and the science and ethics of such communication'. She begins with a brief history of GM tomatoes approved for human consumption in 1994, and a GM tomato marketed in the UK as tomato paste in 1996. These products did not sell well. While there are various reasons for their commercial failure, such as production problems and poor pricing policies, Kolodinsky focuses on the public debate over GM crops and foods. Scientists and food company representatives placed their emphasis on educating the public regarding the science of genetic modification in order to convince consumers that the products were safe to consume. However, this 'science alone' approach to communication failed to address more fundamental concerns that the public had with respect to autonomy, non-maleficence, beneficence and justice, which Kolodinsky argues are key to an effective understanding of the science of communication. Thus, ethical tensions arise when the communication of science does not align with the science of communication. Kolodinsky concludes the chapter with a discussion of how to align better science and communication in order to resolve ethical tensions arising from the use of agricultural biotechnology.

The chapter by Jones, entitled 'Against the (GM) Grain: Ethical Tensions and Agrobiotechnology Activism in the USA', examines the history of public discourse and anti-biotechnology political activism in the USA. Jones presents two cases studies of public dissent in the USA: Occupy The Farm and March Against Monsanto. The case studies show that the early framing of GM foods in terms of harm and risk failed to address deep-seated and persistent concerns about agricultural biotechnology. Simply stated, by attempting to maintain a narrow discourse on health and safety, developers of agricultural biotechnology failed to recognize and appreciate the numerous ethical tensions that agricultural biotechnology created. This provided space for opponents of biotechnology to wage their war in the public arena. The chapter complements Kolodinsky's contribution by making 'it clear that ethical tensions (e.g. community food sovereignty) and cultural anxieties (e.g. ontologies of the natural world and the human place in it) are increasingly important in the political imaginary of emerging counter-publics'.

Aerni's chapter, entitled 'The Use and Abuse of the Term "GMO" in the "Common Weal Rhetoric" Against the Application of Modern Biotechnology in Agriculture', also considers the nature and content of public discourses over agricultural biotechnology but focuses instead on the motives, agendas and 'opportunistic behavior' of its opponents. Opponents of agricultural biotechnology try to position themselves as voices of the public interest, hence their use of 'common weal' or public interest 'rhetoric'. But this creates an ethical tension between their interest in affecting political and economic change and the actual interests of the general public. Moreover, the manner by which opponents of GM technology engage in public narratives raises ethical concerns. Aerni explains how opponents use the term 'genetically modified organism', or 'GMO', as 'a convenient proxy for everything that is believed to be bad about economic and technological change in agriculture'. In this way, they inappropriately make complex social and environmental problems appear as a simple dichotomy of 'people versus profit', such that anything involving GM technology is viewed as bad and anything GM-free is considered good. For example, companies that endorse the use of GMOs in agriculture risk 'becoming a target of symbolic protest'. Moreover, opponents of agricultural biotechnology are not always consistent in what they support and what they reject. For instance, Aerni notes that Europeans opposed to GMOs regularly consume bread made from wheat developed through a non-GM 'mutagenesis' process that contains 'more transgenic material than any currently approved GMO variety'. Aerni then presents two case studies – one regarding the development of GM regulation in Switzerland and the other about the Heubuch Report to the EU Parliamentary Commission on Development – to show that an attempt at a 'common weal rhetoric' does not always advance the common public interest and may actually work against it.

The chapter by Ankeny, Bray and McKinley, entitled 'Collaborating with the Enemy? A View from Down Under on GM Research Partnerships', explores the distribution of funding and approvals for GM crops in Australia in order to determine if private interests dominate public research in agricultural biotechnology and if public interests can be served by collaborations with private organizations. For example, consider the question of whether funding from industry for scientific research in public universities creates conflicts of interests between researchers and funding partners. More generally, does funding from private organizations conflict with the public interest? While other writers raise concerns about the potential conflicts of interest inherent in public–private partnerships, the chapter by Ankeny *et al.* presents a nuanced view to this problem. The authors conduct an 'analysis of applications to the Australian regulatory authority for intentional release of a genetically modified organism (GMO) to explore the actual distribution of public, private and other forms of funding underlying the research'. While there are some patterns in research on specific GM crop traits – private entities seem to focus on certain commercially important traits like herbicide tolerance and insect resistance – it does not appear to be the case that private

interests dominate research on agricultural biotechnology. Moreover, the mix of partnerships is complex. Many agricultural GM projects involve large multinationals, local start-up firms, non-governmental organizations, and even farmer-associated organizations, suggesting that what counts as 'public benefit' needs to be considered carefully. One implication from their research is that concerns about financial conflicts of interests might be overplayed and that biotechnology researchers can effectively work through this particular ethical tension.

The second section contains three chapters that examine ethical tensions associated with the policy and regulation arena. Windsor's chapter, entitled 'Three Models of Public Opinion and Public Interest for Agricultural Biotechnology: Precautionary, Conventional and Accommodative', provides an effective bridge between the first and second sections of this book. Windsor is interested in the question of whether public opinion and public interest should inform on the direction and speed of new technology developments or whether new technology 'should lead (and thus reshape) public opinion'. Government regulation matters here because 'government lies between business (lobbying) and public opinion (citizen consumers and voters)'. Windsor considers three models of how public opinion is linked to regulatory approaches and reviews examples of each. For example, Windsor shows how European regulatory systems are precautionary and that public opinion leads and biotechnology research follows. Many agriculturally oriented countries are conventional; while not being anti-GMO, public opinion leads over new technology research. In the USA, regulation is accommodative, so that agricultural biotechnology leads and public opinion follows. Windsor explains how ethical tensions are affected by each regulatory model, and he suggests that regulatory models can change, by, for instance, moving from precautionary to conventional to accommodative.

In the chapter entitled 'Genetically Modified Organisms in Food: Ethical Tensions and the Labeling Initiative', Strauss is concerned about the adequacy of the regulatory system in the USA to manage the complexity of issues relating to agricultural biotechnology. As evidence, she focuses on the lack of regulations requiring the labeling of foods containing GM ingredients. In her view, consumers 'have the fundamental right to know what they are buying and eating before making a purchasing decision', and the 'reluctance of the US government to establish a rigorous labeling scheme for the treatment of GM crops and GMOs in food' violates that right. The ethical tension here arises when a technology has the potential to cause harm or affect the health and safety of consumers. The problem is one of effectively managing the interests of consumers to be informed and the question of what information can or should be provided to them, and how that communication should occur. A related problem is one of balancing the interests of consumers with the interests of business firms. Strauss believes corporate interests are given too much weight in this calculus. In the case of agricultural biotechnology, this is because the question of informed consent involves more than the issue of potential risk, since 'many consumers make food choices based on religious, ethical and environmental considerations'. Strauss examines legislative efforts to require labeling of GM foods and proposes a framework for addressing the ethical tensions relating to informed consent and labeling.

Kolady and Srivastava begin their chapter, entitled 'Ethical Tensions in Regulation of Agricultural Biotechnology and their Impact on Policy Outcomes: Evidence from the USA and India', with the premise that different regulatory systems produce different types of ethical tensions and concerns, particularly regarding the regulation of agricultural biotechnology. They demonstrate these differences through a comparative study of the regulatory systems in the USA and India and comment on how these differences affect regulatory policy outcomes. For example, because the US regulatory system is relatively centralized (operating through three main federal regulatory agencies), and because it provides a mechanism for public comment, the system 'is less susceptible to external political influence' than the system in India, which consists of a 'complicated hierarchical structure with statutory bodies at state and federal levels' and provides little opportunity for public comments. Hence, regulatory decision making in the USA is effective and predictable, at least compared with the system in India, which is 'complex and uncertain'. However, the centralized regulatory approach in the USA creates a unique ethical tension that is less evident in India. While social and economic concerns can be raised in regulatory decision making in the USA, they rarely make their way into final rules, which largely reflect risk-based concerns about potential harm to human health and the environment. In contrast, social justice concerns can 'become dominant in the policy discourse and influence the regulatory outcomes' in India. In other words, regulatory systems can instigate an ethical tension between support for health and safety and support for social justice.

Two papers comprise the third section dealing with the technological fix arena. Scott's chapter is entitled 'Technological Pragmatism: Navigating the Ethical Tensions Created by Agricultural Biotechnology'. Scott examines 'a polarizing conflict between two opposing philosophies of technology, which can be labeled technological optimism and technological pessimism'. However, while most commentators in the agricultural biotechnology debate can be considered either technology optimists or pessimists, most people are somewhere in the 'muddled middle'. Thus, an important ethical tension arises between what is debated by vested interests and what concerns the general public. Scott frames his discussion within the context of the technological fix criticism. After describing the cultural and historical evolution of optimistic and pessimistic perspectives relating to agricultural biotechnology and their associated ethical tensions, Scott describes a middle approach of technological pragmatism. According to Scott, 'technological pragmatism is capable of providing the conceptual tools to navigate the thicket of ethical tensions created by agricultural biotechnology' because it focuses on 'correcting the defects of the technological fix strategy by including social and ecological factors that contribute to problems and including contributing to more just social arraignments as a goal'. In other words, a pragmatic approach does not turn a blind eye to ethical tensions, but rather seeks to acknowledge and then mitigate those challenges.

In the chapter entitled 'Absolute Hogwash: Assemblage and the New Breed of Animal Biotechnology', MacDonald uses the case of the Enviropig to examine 'why ethical tensions in the food supply must be considered, but also why their form is ever-evolving in light of new technologies'. The Enviropig was genetically engineered with a gene from a mouse to digest phosphorus in its food supply, resulting in the pig excreting significantly less phosphorus than existing pigs, which makes it safer for the environment. There are several ethical tensions relating to the introduction of this type of technology. One comes when something we used to think of as the opposite of technology – a living organism – is now viewed as something technological. According to MacDonald, agricultural biotechnology 'has permitted the rupturing of the very order and role of biology'. Another comes from considering how a small change in a part produces major changes in a whole system. By replacing a tiny part of the hog's genome in order to solve an environmental problem, scientists create a dilemma for consumers who must decide whether they want to consume conventional but environmentally harmful pork or the GM variety that is more environmentally friendly but contains a mouse gene. MacDonald concludes with a discussion of the ethical tension that arises when we accept large-scale animal agriculture and the problems associated with it but are unwilling to accept tiny changes in a food animal's genome.

The fourth section consists of two chapters that address issues relating to the arena of new technologies and that consider specifically the CRISPR technology. The chapter by Pirscher, Bartkowski, Theesfeld and Timaeus, entitled 'Nature-identical Outcomes, Artificial Processes: Governance of CRISPR/Cas Genome Editing as an Ethical Challenge', examines in detail ethical tensions arising from the CRISPR technology described above. Pirscher *et al.* note that current uses of CRISPR in agriculture are challenging because they '[do] not leave detectable traces of genetic engineering (especially foreign DNA snippets) in the resulting organism, thus its results effectively cannot be identified as genetically modified organisms (GMOs)'. In other words, traditional debates about how GM plants and animals are 'unnatural' and cross species boundaries or pose risks to the environment break down in the case of 'cisgenic' (or within species) modification in contrast with 'transgenic' modification. Cisgenic modification also raises important questions about regulation. Should regulation specifically, and ethical debates generally, focus on the process by which genetic modifications occur (e.g. in the laboratory) or should it address the final product (e.g. the Enviropig, which contains the gene of a mouse)? In other words, the CRISPR technology shifts the debate more heavily on processes from products, and this shift has implications for understanding ethical tensions arising from new technology.

In their chapter entitled 'New Technology, Cognitive Bias and Ethical Tensions in Entrepreneurial Commercialization: The Case of CRISPR', Ng and James also consider CRISPR but in the context of how new technology creates cognitive challenges for innovators. Ng and James are interested in the ethical judgments of innovator-entrepreneurs when making decisions about the commercialization of new technologies. This concern is not based merely on economic considerations, where innovator-entrepreneurs have incentives to pursue their own self-interests instead of the interests of broader stakeholders. Rather, innovator-entrepreneurs operate in complex and novel decision settings, which 'can result in "egocentric" and "myopic" biases that impair an entrepreneur's ethical judgments with respect to broader societal groups'. In other words, innovation can make individuals vulnerable to an ethical blindness that makes it particularly difficult for them to take into consideration broader rather than narrow interests. Ng and James also note an interesting ethical tension 'between the cognitive and personality traits needed to commercialize [new technology, such as] CRISPR/Cas9 and those that affect the ability of entrepreneurial innovators to identify with the ethical concerns of indirect stakeholders'. Consequently, Ng and James argue that developers and adopters of new technologies need to be particularly vigilant in taking into consideration the interests and perspectives of society and other stakeholders.

The final section contains one chapter that provides insights into mediating ethical tensions arising from new technology adoption. This chapter, by Valdivia, James and Quiroz, is entitled 'New Technology, Ethical Tensions and the Mediating Role of Translational Research'. If new technology creates a conflict of interests, values or rights, then the use of translational research methods in the development or adoption decisions of new technology might be a way of alleviating some of these ethical tensions. The reason is because a hallmark feature of translational research is the facilitation of two-way communication among relevant stakeholders. When done well, translational research processes can improve trust and alleviate concerns that have the potential of limiting perceived benefits of new technology. According to the authors, the 'approach has addressed many ethical tensions between different stakeholder groups that belong to the practice of farming' and has the benefit that it 'facilitates or enhances the voice of smallholder farmers'. The authors consider two specific case studies, GM cassava in Kenya and the use of unmanned aerial vehicles in East Africa, to show how translational research can mitigate ethical challenges from new technologies. They are able to demonstrate that 'while translational research will not solve all ethical tensions that new technology creates, it can be effective in mitigating important challenges that arise from conflicting, or misunderstood, interests and concerns of stakeholders'.

Lessons for Understanding Ethical Tensions Arising from the Introduction of New Technology

It is not uncommon to hear or read the expression 'technology is not neutral'. What this means is that technology has social, economic and environmental implications.

One way to examine these implications is through the ethical tensions that new technology produces. An ethical tension occurs when a technology creates or has the possibility of creating a conflict of interests, values or rights. If it is true that technology is not neutral, then any and every new technology will result in at least some type of ethical tension. Sometimes the ethical tensions of technologies are obvious, but often they are not. Thus, one lesson from this book is that we need to be particularly mindful about the ethical challenges that new technology creates.

Knowing where and how to look for ethical tensions is the first step in understanding them. To this end, this books considers four ways or arenas within which new technology creates ethical tensions. In order to explore these tensions as well as ideas for resolving or mitigating them, the contributions in this book focus on the case of agricultural biotechnology. Although agricultural biotechnology is not new, this book can inform on the ethical tensions and social, economic and environmental implications of new technology. If agricultural biotechnology is helpful in this regard, then another key takeaway is that there is no shortage of ethical tensions that agricultural biotechnology produces.

References

Biotechnology Industry Organization (BIO) (2006) Guide to biotechnology. Available at: http://www.bio-nica.info/biblioteca/BIO2006BiotechGuide.pdf (accessed 9 February 2018).

Cyranoski, D. and Reardon, S. (2015) Embryo editing sparks epic debate. *Nature* 520, 593–595. doi:10.1038/520593a

Daniels, M. (2017) Avoiding GMOs isn't just anti-science. It's immoral. Opinions, *The Washington Post*, 27 December. Available at: https://www.washingtonpost.com/opinions/avoiding-gmos-isnt-just-anti-science-its-immoral/2017/12/27/fc773022-ea83-11e7-b698-91d4e35920a3_story.html?utm_term=.91438237c9a2 (accessed 22 January 2018).

David, K. and Thompson, P.B. (2008) *What Can Nanotechnology Learn from Biotechnology? Social and Ethical Lessons for Nanoscience from the Debate over Agrifood Biotechnology and GMOs*. Academic Press, Burlington, Massachusetts.

Holland, A. and Johnson, A. (eds) (1998) *Animal Biotechnology and Ethics*. Springer, Dordrecht, Netherlands.

Jorgensen, E. (2016) What you need to know about CRISPR. TED talk, June. Available at: https://www.ted.com/talks/ellen_jorgensen_what_you_need_to_know_about_crispr (accessed 22 January 2018).

Ledford, H. (2015) CRISPR, the disruptor. *Nature New* 522, 20–24. doi:10.1038/522020a

National Academies of Sciences, Engineering, and Medicine (NAS) (2016) *Genetically Engineered Crops: Experiences and Prospects*. NAS, Washington, DC. doi:10.17226/23395

Ong, A. and Collier, S.J. (eds) (2005) *Global Assemblages: Technology, Politics, and Ethics as Anthropological Problems*. Blackwell Publishing, Malden, Massachusetts.

Qaim, M. (2016) *Genetically Modified Crops and Agricultural Development*. Palgrave Macmillan, New York.

Schaub, P., Al-Babili, S., Drake, R. and Beyer, P. (2005) Why is Golden Rice golden (yellow) instead of red? *Plant Physiology* 138, 441–450. doi:10.1104/pp.104.057927

Scott, D. (2011) The technological fix criticisms and the agricultural biotechnology debate. *Journal of Agricultural and Environmental Ethics* 24, 207–226. doi:10.1007/s10806-010-9253-7

Stein, A. (2015) The poor, malnutrition, biofortification, and biotechnology. In Herring, R. (ed.) *The Oxford Handbook of Food, Politics, and Society*. Oxford University Press, Oxford, pp. 149–180.

Stewart, P.A. and Knight, A.J. (2005) Trends affecting the next generation of U.S. agricultural biotechnology: Politics, policy, and plant-made pharmaceuticals. *Technological Forecasting & Social Change* 7, 521–534. doi:10.1016/j.techfore.2004.03.001

Twine, R. (2010) *Animals as Biotechnology: Ethics, Sustainability and Critical Animal Studies*. Earthscan, New York.

World Health Organization (2016) Biofortification of staple crops. Available at: http://www.who.int/elena/titles/biofortification/en/ (accessed 22 January 2018).

Ye, W.D., Al-Babili, S., Kloti, A., Zhang, J., Lucca, P., Beyer, P. and Potrykus, I. (2000) Engineering the provitamin A (beta-carotene) biosynthetic pathway into (carotenoid-free) rice endosperm. *Science* 287, 303–305. doi:10.1126/science.287.5451.303

1 Ethical Tensions from a 'Science Alone' Approach in Communicating Genetic Engineering Science to Consumers

Jane Kolodinsky*

Department of Community Development and Applied Economics, University of Vermont, Burlington, Vermont, USA

Introduction

Communication experts have known for years that top-down approaches to 'teaching' the public about the benefits of technological advances in agriculture, including genetic engineering (GE), do not change consumer opinions about these technologies (Wohl, 1998; Hails and Kinderlerer, 2003; Hansen *et al.*, 2003; Landrum and Hallman, 2017). Nevertheless, there remains a view that a focus on educating consumers about science will help gain public support for GE (Blancke *et al.*, 2015). To this end, in 2017 the FDA allocated US$3,000,000 for a campaign to publish and distribute science-based educational materials to make sure that people understand the benefits of GE (Dewey, 2017). When the science of GE ignores the science of communication, it seems that science is at odds with itself. And when the science of GE ignores important ethical principles of communication, ethical tensions will arise.

This chapter identifies and examines the ethical tension between the need to communicate the risks and benefits of science and technology to the public and the science and ethics of such communication.

It begins with a historical background of the commercialization of GE foods and places the ethical debate about GE foods in the bioethics rubric of autonomy, non-malfeasance, beneficence and justice. Ethical tensions about GE foods go beyond safety and include difficult to quantify issues, such as interference with the natural order of things, religion, freedom of choice, environmental concern, trust, risk, cultural identity, equity, fairness, consent and self-determination. Through an analysis of the literature, including scientific, gray and popular, this chapter shows that policies promoting autonomy can appease ethical concerns of consumers by providing choice in the marketplace. While public debate will likely continue, informed individual consumer choice can help shape the future of GE foods.

Historical Perspective: The Flavr Savr and Zeneca Tomatoes

Early in the introduction to GE into the mainstream food supply, industry knew there would be a problem, given early opposition from organizations such as the Environmental Defense Fund, The Pure Food

* E-mail: jane.kolodinsky@uvm.edu

Campaign and the New York State Consumer Protection Board in the USA, and consumer protests and unfavorable media coverage in the United Kingdom (Leary, 1994; Elderidge, 2003). Confrontation, controversy, irrational and anxiety are words neither consumer nor seller wants associated with their product (see, for example, Buzalka, 2000; Marris, 2001). These tension-causing words erupted during the early introduction of GE products into the human food supply. The first product was a tomato called the Flavr Savr. Words associated with GE have not changed in the past 25 years. Consideration of the introduction of GE tomatoes is informative.

The Flavr Savr tomato was created and commercialized by Calgene, a California tech start-up company (CERA, 2015). It was approved by the US Department of Agriculture (USDA) and went on the market in 1994. Calgene was a self-professed 'techie' company and not a farming business, and the transformation from laboratory to production and distribution did not go well. The Flavr Savr was grown in Mexico, poorly packed and shipped, and arrived at supermarkets ruined. The price charged was often twice as high as that for a conventional tomato. Even when it looked good when shipped correctly, the tomato did not have a taste worth twice the price. So, Calgene was sold to Monsanto in 1996 and the tomato was quickly pulled from the market (Associated Press, 1996). The Monsanto website FAQ provided an answer as to why the Flavr Savr is no longer available: 'weak harvests and costly shipping methods; the tomato was unprofitable' (Monsanto, no date). Other sources indicate that the Flavr Savr was discontinued because Monsanto had loftier goals for genetic engineering than tomatoes; they aimed to be a biotechnology company (Center for Food Safety, 2005). Importantly, the Flavr Savr was voluntarily labeled. By pulling the product from the market, labeling became a non-issue. If GE had the goal of feeding the world, selling a tomato at twice the price of a conventional tomato did not fit that aim.

In the United Kingdom, a similar GE development was in the works at Zeneca, initiated by Dr Don Grierson, then a plant geneticist at the University of Nottingham. Zeneca obtained approval for a GE tomato in August 1994 (Bruening and Lyons, 2000). Zeneca's idea was to produce a processed tomato paste sold in larger cans and the same price as conventional tomato paste. It was introduced in 1996 and clearly labeled GE.

Grierson noted that the tomato paste version was a better product because it was labeled so that consumers could make their own choice, and it met one of the goals of GE technology – to provide more food at a lower price. This was in contrast to the relatively higher priced Flavr Savr (Goddard, 1998). According to Grierson, 'The key thing is that the product is labelled. … This is not a scientific issue. It is not a safety issue. It is just a question of recognising people's individuality, and their concerns, and their right to know' (Goddard, 1998, p. 1). These words could be written today, as the first mandatory labeling of GE foods in the USA was promulgated based on consumers' right to know (Vermont General Assembly, 2014). Additionally, The National Bioengineered Food Disclosure Law S764 was passed in 2016, charging the Agricultural Marketing Service of the USDA to develop labeling standards within two years (USDA, no date).

The Zeneca tomato initially sold well in Safeway and Sainsbury stores, but in 1998 sales began to decrease and the product was pulled from supermarket shelves. Supermarket giant Tesco, which never stocked the GE puree, stated the puree did not offer any additional benefits to customers compared with conventional puree (BBC News, 1996). Interestingly, a major tenet of marketing and advertising is 'give me benefits, not attributes' (Belch and Belch, 2012). The benefits of the tomato did not go to the consumer, but rather to the supply chain, as the 'farmer has a longer window for delivery, there is less mold damage, the tomatoes are easier to transport and they are better for processing' (BBC News, 1996, n.p.).

Opponents of GE cited flawed research findings about potential health dangers of GE potatoes (Bruening and Lyons, 2000). But others blame Árpád Pusztai, until then a world-renowned biochemist and nutritionist. Pusztai was vilified by other scientists for publishing a study (Ewen and Pusztai, 1999) and talking to the media about the possibility of negative health effects of GE. He wasn't the last (see Krimsky, 2015). Reading *A United Kingdom Parliament Report* (Science and Technology Committee, 1999) on genetically modified foods, it is clear that there were several possible reasons for the tomato's failure, including consumer choice, the costs of knowing which tomatoes were GE (separation of product), the media, lobbying groups and consumer perceptions. Pusztai was also mentioned. Nevertheless, the confrontation of scientist against scientist cost Pusztai his position at the Rowett Institute in Scotland and no doubt contributed to the controversy that was brewing in consumers' eyes with regard to GE food.

The message from the stories of the Flavr Savr and Zeneca tomatoes is that the development of GE technologies in agriculture is complex and controversial. Moreover, there continues to be information and misinformation surrounding the health, safety and other issues related to the diffusion of GE into the food supply. Thus, it is difficult to determine who and what is to blame for the less than universal acceptance of GE technology into the US food supply (Kolodinsky, 2007).

Ethical Issues Go Beyond Safety

The Flavr Savr and Zeneca stories are important because they provide contexts for a discussion about the ethical concerns consumers face in the USA and elsewhere. In addition to words found in the media, including controversy, confrontation and anxiety, we can add interference with the natural order of things, religion, freedom of choice, environmental concern, trust, risk, cultural identity, equity, fairness, consent, autonomy, self-determination and perceptions of safety (Thompson, 1997; Kolodinsky et al., 1998; Wohl, 1998; Hansen et al., 2003; Lang and Hallman, 2005; Dryzek et al., 2009; Gupta et al., 2011; Hepting et al., 2014; Fischer et al., 2015; Landrum and Hallman, 2017). In short, there are issues beyond safety that people

are concerned about and this is seen in a wide variety of disciplines and over a long period of time.

Slovic (1987) offers a perspective about risk that remains relevant today. Using factor analysis, he developed two data-driven axes with regard to risk: known versus unknown, and acceptable versus dreaded. DNA technology landed in the combination of unknown and dreaded, above every other 'risky' activity, drug or technology. This research was conducted ten years before the introduction of the first GE food on the market. Slovic noted several characteristics of the unknown and dreaded combination: 'not observable, unknown to those exposed, effect delayed, new risk, risks unknown to science, uncontrollable, dreaded, global catastrophic, consequences fatal, not equitable, risk to future generations, not easily reduced, risk increasing, and involuntary' (p. 283).

A superficial understanding of Slovic would suggest that scientists and proponents of GE are doing it right – they are communicating to the public that GE is safe and not dreaded, and that risks are knowable and acceptable. But this 'science alone' approach does not account for many other ethical concerns that consumers have, some of which are highlighted in the paragraph opening this section of the chapter. For this reason, a bioethics ethics framework is useful (Beauchamp and Childress, 2001; Hepting et al., 2014). Mepham (2000) adapted bioethics into a framework for the analysis of novel foods. Mepham's (2000) ethical matrix is useful in the discussion of consumer's ethical tensions with regard to GE technologies. It takes abstract notions and maps their relationship to actual behaviors. The principles include: autonomy, non-maleficence, beneficence and justice.

Autonomy acknowledges a consumer's right to make choices and take actions based on values. *Non-maleficence* means 'do no harm'; the idea is that GE technology should not inflict harm intentionally. *Beneficence* relates to acts done to benefit others and to contribute to societal welfare. *Justice* considers equity and the fair distribution of benefits and burdens.

Some readers of these definitions may say 'See, biotechnological advances meet all the ethical principles', while others might say 'See, there is clearly a problem with introducing biotechnological advances to the marketplace'. Why are there opposing perspectives of the science of genetic engineering when viewed from a lens of ethics?

Rollin (2014, p. 511) calls the interplay among genetic engineering, science and ethics a perfect storm, words used to describe 'untoward situations in all areas of life or society where the result is more problematic than the sum of its parts'. He blames both scientists and consumers. In the USA, people are generally scientifically illiterate, and scientists have a poor understanding of ethics and principles of effective communication. Since the removal of the Flavr Savr and Zeneca tomatoes from the marketplace, the US government, industry and many (but not all) scientists continue to stress, at the expense of other issues, the need to educate consumers about the science and safety of GE foods. Yet this focus has not appeased consumer concerns nor increased consumer acceptance of GE technologies. Known as the 'science alone' model, we as a society leave it to the experts to assess probabilistic risk and then to inform the public about it (Levidow and Carr, 1997; Grove-White and Szerszynski, 1992). But there are problems with this approach when viewed against the ethical principles of autonomy, non-maleficence, beneficence and justice, as explained below.

Autonomy

Autonomy has received the most attention as it relates to consumer concerns. The consumer behavior literature has two core principles that include value=benefits/price and 'show me benefits, not characteristics' (Belch and Belch, 2012). Without autonomy, none of these can become choice criteria for consumers. When consumers cannot identify which products contain GE ingredients or were produced or partially produced using GE, they are unable to make a choice that

may relate to any other of the four values. While the economic literature about GE has focused on consumer willingness to pay as a proxy for preference (e.g. Costa-Font et al., 2008), many evaluative criteria that denote 'price' for consumers are not purely monetary, but also express social or psychological value. Monetary value is a function of the price paid and is relative to an offering's perceived worth. Functional value is what an offer does; it's the solution an offer provides to the customer regardless of monetary price. Social value is the extent to which owning a product or engaging in a service allows the consumer to connect with others. Psychological value is the extent to which a product allows consumers to express themselves and is more personal. These and other food values can be tied to a person's identity. For example, Lusk and Briggeman (2009) outline 11 food values, all of which can or will be tied to GE technology in one way or another: naturalness, taste, price, safety, convenience, nutrition, tradition, origin, fairness, appearance and environmental impact.

The biotechnology literature does not directly link consumer autonomy with choices that are of value to consumers. Instead, it links autonomy with the *inability* to choose due to lack of disclosure, because, for instance, GE foods are generally not labeled in the USA. Dieterle (2016) goes deeper into the labeling discussion to the notion of informed consent, which includes both disclosure and consent. In this arena, the 'science alone' approach becomes irrelevant. Instead, the consumer makes their decisions based on their individual assessment of GE. This is in contrast to the assertion that GE labeling is the opposite of autonomy (Carter and Gruère, 2003). It assumes that labels will increase dread; thus, products produced using GE will be eliminated from the marketplace and consumer autonomy will be reduced. Of course, this is an empirical question. In the USA there are signs from studies that simulate a labeling environment suggesting that dread does not increase and labels do convey information (Costanigro and Lusk, 2014; Liaukonyte et al., 2015; Kolodinsky, 2008, 2015). Evidence from the only US state that mandated GE labels in an actual marketplace shows that labeling may actually increase support for GE technology (Kolodinsky, 2017). And, even without mandatory labeling, there has been an emergence of voluntary labeling and signs that firms are producing products with and without GE ingredients. As the USA develops the standards for national mandatory labeling, the story will unfold.

Non-malfeasance

At the forefront of most people's minds is the safety of GE foods, with no adverse health effects. The 'science alone' model is the basis for this. In the most comprehensive review to date by the National Academies of Sciences, the chapter devoted to safety concludes:

> However, the research that has been conducted in studies with animals and on chemical composition of GE foods reveals no differences that would implicate a higher risk to human health from eating GE foods than from eating their non-GE counterparts … The committee could not find persuasive evidence of adverse health effects directly attributable to consumption of GE foods. (2016, p. 66)

The 'however' beginning the statement leads the curious reader to wonder why it is needed. The report stated that no adverse health effects have been found using current methods, but also that the scientific community must find new methods of testing. There are 27 findings in the chapter related to negative health effects. About one third of the statements relate to facts and current and emerging technologies (such as gene editing). About one third of the statements specifically say that GE foods have not been found to be related to kidney disease, obesity, celiac disease, allergies, autism or cancer. However, about one third of the statements indicate that current methods of testing are inadequate, association does not imply causation, significant differences have been found but are not out of a normal range, significant differences are small, more studies are needed and that there is disagreement among scientists. In other words, we just

don't know everything now about the safety of GE food, given current evidence. Perhaps this is the reason the word 'however' is in the conclusion of the report. For readers wanting evidence counter to a 'scientific consensus' on the safety of GE technology, Krimsky (2015) and Hilbeck *et al.* (2015) are places to begin.

So why do consumers continue to question the safety of GE foods? People prefer certainty and not probabilistic calculations of risk. The National Academies of Sciences report is not the only publication that has noted we may not have the right tests for new technologies (e.g. Bawa and Anilakumar, 2013; Klümper and Qaim, 2014; Krimsky, 2015). The National Academies of Sciences report also states that 'change or risk could be something that has not even been considered, so the only effective testing is of the whole food itself' (2016, p. 178) and arrives at summary findings other than safety that exacerbate notions of uncertainty, including:

- The nation-wide data on maize, cotton and soybean in the USA do not show a significant signature of genetic-engineering technology on the rate of yield increase (p. 102).
- Both for insect pests and weeds, there is evidence that some species have increased in abundance as intermediate resistance and high resistance crops have become widely planted (p. 136).
- Although gene flow has occurred, no examples have demonstrated an adverse environmental effect of gene flow from a GE crop to a wild, related plant species (p. 152).
- Both GE crops and the percentage of cropping area farmed with no-till and reduced-till practices have increased over the last two decades. However, cause and effect are difficult to determine (p. 154).
- The economic effects of mandatory labeling of GE food at the consumer level are uncertain (p. 306).

The case of the herbicide glyphosate, which is not a GE product but is used in combination with some GE seeds, is also illustrative. A recent article on the safety of glyphosate concludes no scientific association between glyphosate and cancer in a large study of certified pesticide applicators

(Andreotti *et al.*, 2018). However, a companion article to Andreotti *et al.* adds an important caveat. Comparing the study findings to historical research on benzene, Ward notes:

> Evaluating the potential for glyphosate exposure to increase cancer risks in humans is important due to its widespread and increasing use in the United States and globally and indications of potential carcinogenicity from toxicologic and epidemiologic studies. Epidemiologic studies have inherent limitations with respect to cancer prevention as they generally detect elevated cancer incidence and mortality cancer hazard decades after carcinogen exposure begins. (2018, p. 2)

Taleb's (2010) and Tversky and Kahneman's (1974) narratives about uncertainty and risk are applicable here. The average citizen is scientifically illiterate and does not understand probabilistic risk assessments in common discourse (Rollin, 2014; Blancke *et al.*, 2015). But it is not the only discourse. Scientific illiteracy leads people to make 'irrational' decisions (Tversky and Kahneman, 1974). In contrast, Taleb (2010) asserts that instinct and 'gut feelings' can also lead to good decisions. Thus, it is easy to see why tensions persist. Careful reading of the literature can cast some doubt on the safety and efficacy of current GE crops. Consumers who prefer certainty or have uneasy feelings about these technologies have not been appeased by information provided to date.

There is also debate about malfeasance on the part of Monsanto that does not deal directly with the safety of GE, but rather with the process by which GE technologies came to market. An early product produced using GE is rBST (recombinant bovine somatatropin). Monsanto marketed (as early as 1991) rBST to farmers before it gained federal approval and was determined safe to use on dairy herds. But reports obtained in 2009 via the Freedom of Information Act showed the company knew that the drug caused more health and mastitis incidents and reproductive problems in cows than in a control herd, that intramuscular injection was unacceptable and that there was an increase in total cows affected and total days observed for hock abnormalities (FDA,

2009). Simply stated, Monsanto sold a product that it knew to be unsafe for dairy cows.

Two other examples are illustrative. The first is probable coordination between Monsanto and the EPA with regard to the possible carcinogenic effects of glyphosate and the second is Monsanto possibly exerting industry influence through ghostwriting. An email exchange between Monsanto executives Dan Jenkins (no date) and William Heydens (no date) explicitly suggests a Monsanto/EPA exchange:

> NOW THE QUESTION – What are your thoughts on approaching EPA and having a conversation, probably a generic one, about what area they see as most problematic...?

In another email Heydens (no date) indicates he has edited an expert panel summary, later published as Williams *et al.* (2016), finding no cancer relationship with glyphosate:

> OK, I have gone through the entire document and indicated what I think should stay, what can go, and in a couple spots I did a little editing. I took a crack at adding ... see what you think....

While the emails indicate a different journal and set of authors, Dr Gary Williams ultimately was first author on the publication. There is a manuscript trail of Williams citing Monsanto researchers and specifically Heydens in an earlier publication (see Williams *et al.*, 2000). Waldman *et al.* (2017) have a more popular press take on the issue of ghostwriting.

Another example about malfeasance on the part of Monsanto may be part of an unfolding story of Xtend soybeans. Brad Williams, a Missouri farmer, noted 'I'm a fan of Monsanto. I've bought a lot of their products. I can't wrap my mind around the fact that there would be some kind of evil nefarious plot to put a defective product out there intentionally' (Hakim, 2017, n.p.). The new soybean seed was genetically modified to resist both glyphosate and dicamba due to a growing pigweed problem. Monsanto began selling dicamba to farmers before EPA approval of a new way to spray dicamba, which is toxic not only to non-resistant varieties of crops, but also to woody plants and trees (Mericle, 2017; Nosowitz, 2017). The only alternative to waiting for approval would have been 'to not sell a single soybean in the United States', Monsanto Vice President of Global Strategy Scott Partridge told Reuters in an interview (Flitter, 2017).

A report by University of Missouri scientists estimates that '2,708 dicamba-related injury cases [are] currently under investigation by various state departments of agriculture around the U.S., and that there were approximately 3.6 million acres of soybean that were injured by off-site movement of dicamba at some point during 2017' (Bradley, 2017, n.p.). There is evidence that Monsanto knew of potential issues with selling seed, knowing farmers could not use the very herbicide the seeds were meant to resist. In this case the sale of Xtend soybeans did more than harm the revenue potential of individual farmers; it impacted millions of acres of soybeans and thousands of growers. As this story is still unfolding, the extent of damage and the remedy needed are still unknown. Regardless, it points to a strong possibility of intentional malfeasance on the part of a major company and no doubt will increase the ethical concerns of consumers about GE technology.

Can a 'science alone' approach be the best path forward when information after-the-fact becomes available that discounts previous scientific statements about safety? And what about cases where companies move forward with the marketing of products before science has been allowed to weigh in on their safety? These questions imply that even an iterative 'science alone' approach is not sufficient in addressing consumer concerns about GE technology. Yet, communication that GM technology is safe remains the message of the pro-GE community, even while the message of the anti-GE community is for science communication to reflect other values of importance.

Beneficence

According to a plant pathologist and geneticist at the University of California at Davis,

Dr Pamela Ronald, beneficence means a focus on tangible benefits:

> All this arguing about what's genetically modified is a big distraction from the really important goals ... We need to produce safe and nutritious food that consumers can afford and farmers can make a profit from. And we need agricultural practices that enhance soil fertility and crop biodiversity, use land and water efficiently, reduce use of toxic compounds, reduce erosion, and sequester carbon. I think most everyone agrees on those general principles. (Berlin, 2015, n.p.)

Dr Ronald is not touting GE technology as the only way forward, as has the popular press and a few stalwart scientists. Rather, she raises the issue of what benefits consumers receive from GE foods. The case of Golden Rice is informative. Golden Rice was introduced as a silver bullet solution to world hunger (Nash, 2000). Monsanto and other biotechnology companies launched a US$50 million marketing campaign, including US$32 million in TV and print advertising, asserting that 'biotech foods could help end world hunger' (Robbins, 2011, n.p.). But Golden Rice is still not available for use. Stone and Glover (2017) found that the anti-GE movement was not the impediment to commercialization. Dubock (2014) blames what he calls an antiquated protocol for biosafety to be the reason. Ethical tensions emerge and remain because science, industry, the public, governments and NGOs cannot arrive on the same page. The result is that the consuming public is left to self-determine benefits and risks, often from unreliable sources.

There are so many websites that are propaganda for both sides of the GE beneficence story. In fact, a Google search (on 20 November 2017) with the search line 'will GMOs end world hunger' revealed more than 3,000,000 hits. Searching using the term 'GMOs will end world hunger' revealed 286,000 hits and 'why GMOs won't end world hunger' returned 811,000. There are also published articles that take a stand on one or the other side of the debate about whether GE crops are necessary to solve world hunger (Jacobsen et al., 2013; Fedoroff, 2015; Krimsky, 2015).

The quote by Dr Ronald above suggests the need to have a cornucopia of methods to address world hunger and the future of agriculture. The Food and Agriculture Organization of the United Nations (FAO) reports that world hunger is on the rise. In 2016, 815 million people around the world were malnourished (FAO, 2017). The United Nations Sustainable Development goal number 2 is to 'end hunger, achieve food security and improved nutrition and promote sustainable agriculture' (UN, 2017). The FAO reports that the major contributors to hunger include conflict, climate change related shocks and economic slowdowns (FAO, 2017). These issues go beyond planting of GE crops. They impact individual farmers, prices, food availability and trade, even when crops are planted.

The United Nations explicitly talks of resilient sustainable agriculture practices and the maintenance of genetic diversification (UN, 2017). It also speaks of the need for increased agricultural production, which can include the use of GE seed. That said, lack of government investment in all types of agriculture, destabilization of currencies (which increased the price of food), increases in the price of fuel and shifts away from investments in infrastructure for agriculture were all cited as reasons for lack of progress in reaching sustainable development goal number 2 (UN, 2017).

The US government knew that GE was not a magic bullet solution. For example, former US Secretary of Agriculture Dan Glickman said this in a 2001 interview:

> What I saw generically on the pro-biotech side was the attitude that the technology was good, and that it was almost immoral to say that it wasn't good, because it was going to solve the problems of the human race and feed the hungry and clothe the naked. . . . And there was a lot of money that had been invested in this, and if you're against it, you're Luddites, you're stupid. That, frankly, was the side our government was on. . . . You felt like you were almost an alien, disloyal, by trying to present an open-minded view on some of the issues being raised. So I pretty much spouted the rhetoric that everybody else around here spouted; it was written into my speeches. (Lambrecht, 2001, n.p.)

GE technology is one of *many* approaches that can have a positive impact on society. Failing to acknowledge the complexity of the hunger problem by GE proponents and failing to recognize the contribution of GE crops as part of a solution by GE opponents is not due only to 'making a line in the sand'. It is also due to media's necessity to provide soundbites about particular issues when they communicate to a larger audience. Unfortunately, this perpetuates an 'us against them' mentality on both sides of the issue and contributes to the problem of a 'science alone' approach to science communication.

Justice

If autonomy, non-malfeasance and beneficence are less than clear with regard to GE, then justice is by far the most complicated. Justice implies what 'should be', a normative concept with no clear answers. Dryzek *et al.* (2009) use an interesting bi-polar scale that provides insight. On the one hand there is the precautionary approach, taken by much of the world except the USA. According to a United Nations report on the environment and development,

> In order to protect the environment, the precautionary approach shall be widely applied by States according to their capabilities. Where there are threats of serious or irreversible damage, lack of full scientific certainty shall not be used as a reason for postponing cost-effective measures to prevent environmental degradation. (1992, n.p.)

On the other hand is the Promethean view, which places faith in the capacity of humans to manipulate complex systems to their own advantage (Dryzek *et al.*, 2009).

Social impact is a general form of justice. Fischer *et al.* define social impact as 'the consequences to human populations of any public or private actions that alter the ways in which people live, work, pay, relate to one another, organize their needs, and generally cope with society' (2015, p. 8600). The authors note in their review that economic impacts dominate the literature, while access to benefits of GE are related to intellectual property rights and private industry dominance, and that 'well-being' is not well studied. Loss of cultural heritage, loss of agricultural knowledge, loss of seed sharing can be added to the list. The Fischer *et al.* (2015) paper provides an extensive bibliography for readers. These authors fall on the precautionary side of the scale; there is little evidence in the published literature that has accounted for justice-related issues that cannot be monetized.

Fairness is a concept related to justice. Hepting *et al.* (2014) identify issues of fair trade, affordable food, humanity, sustainability and environmental impact under this component of justice. Bennett *et al.* (2013) add intellectual property rights, appropriate roles of the public versus private sector in decision making and who makes decisions about regulation of GE technologies to the list of issues. These authors fall on the Promethean side of the scale, as they explicitly point to the ability of scientists to calculate the risk probabilities of GE with regard to a host of outcomes, including justice.

What does this mean for the public? First, there is a clear split between those who believe in precaution versus those who believe humankind has more control over nature. Add to this the inability to agree on what constitutes well-being for a host of actors, including business, labor, individuals and the environment. Add to this a wide variation in beliefs of how decisions about GE should and are being made. Add to this uncertainty in even the most robust scientific studies of GE on safety and health. It is no wonder there remains polarization, uncertainty and tension about GE in the citizen arena.

Discussion

How do we account for ethical concerns that consumers have regarding GE technologies? Suggested principles of communication and tools from the ethics literature indicate that building trust and dialogue is paramount and relationships are key (Beekman and Brom, 2007; Medlock *et al.*, 2007; Hepting

et al., 2014; Rollin, 2014; Asveld *et al.*, 2015). Top-down approaches to convincing consumers that GE technologies are the answer to world hunger and agricultural productivity have not been convincing. Transparency has not been at the forefront in the development of agricultural applications of GE. Consumers have not been able to observe the benefits of GE, as most benefits have accrued to industry or farmers, and some benefits are waning due to the increase in the development of herbicide resistance in weeds (Fernandez-Cornejo *et al.*, 2014; National Academies of Sciences, 2016). Instead, the pro-GE community has been committed to using a 'science alone' approach. Public consultations have been suggested as a gold standard, but it is difficult to engage large numbers of citizens in this way. Certifications have been suggested, which include using organic certification, of which use of GE ingredients is only one criterion. Seven 'practical' principles have been suggested to address moral and instrumental norms in food risk communication (Modin and Hansson, 2011):

1. Be honest and open;
2. Disclose incentives and conflicts of interest;
3. Take all available relevant knowledge into consideration;
4. When possible, quantify risk;
5. Describe and explain uncertainties;
6. Take all the public's concerns into account; and
7. Take the rights of individuals and groups seriously.

Interestingly, while these points were published in the *Journal of Business Ethics* and were about food safety risk, they are applicable to the GE issue as well. The authors assert that an instrumental justification would be one that shows dishonesty makes it difficult to achieve non-moral goals. Dishonesty will be discovered and make the public suspicious of the business (Modin and Hansson, 2011). This has been shown in the case of GE in the above narrative in the section on non-malfeasance. In three cases that came to light after the fact (rBST and animal health, interference with scientific publication and pre-selling of Xtend soybeans) consumer trust in industry was eroded.

Given controversy, debate and a continued inability to agree on the science of GE, at the very least consumer autonomy may be the cleanest path toward giving consumers choice, which can address their personal assessments of the use of GE in food production. Labeling – voluntary and mandatory – is a viable solution.

By mid-2019, the USA will become the 65th country requiring mandatory labeling of products produced or partially produced using GE or GE ingredients (Just Label It, no date). On 29 July 2016, President Obama signed into law the National Bioengineered Food Disclosure Standard (Public Law No. 114-216), directing the USDA to establish a national standard to disclose certain food products or ingredients that are 'bioengineered'. While the exact method of labeling is currently being determined by the USDA (USDA, no date), consumers will be able to make autonomous decisions about foods produced using GM.

There are other, voluntary labeling arrangements already in place. The Non-GMO Project label

> gives shoppers the assurance that a product has completed a comprehensive third-party verification for compliance with the Non-GMO Project Standard. When it comes to food labeling, third-party certifications are best because they ensure the claim is unbiased, rigorous and transparent. (Non-GMO Project, no date, n.p.)

And in 2015, the FDA issued guidance on voluntary labeling to indicate whether foods are derived from genetically engineered plants and from genetically engineered Atlantic salmon (FDA, 2015a, 2015b).

The US approach to educate the public about the safety of GE has not appeased consumers, nor made any of these issues disappear. A 'science alone' approach has been unable to decrease polarization of consumers in the GE debate. It is unlikely to decrease uncertainty in the future. Therefore, providing consumers with tools that allow them to make their own decisions about GE through purchases in the marketplace may ultimately decide the future of these technologies in the USA, including the balance between food produced or partially produced using GE. By promoting an autonomous public, those who embrace GE technologies, those who do not embrace GE technologies *and* those who do not care either way will be able to advance their own preferences as individuals, given their values. While greater tension may remain at the public level, facilitating consumer choice can buffer the debate.

References

Andreotti, G. *et al.* (2018) Glyphosate use and cancer incidence in the agricultural health study. *JNCI: Journal of the National Cancer Institute* 110, djx233. doi:10.1093/jnci/djx233

Associated Press (1996) Monsanto in deal to acquire Calgene. *New York Times*, 1 August. Available at: http://www.nytimes.com/1996/08/01/business/monsanto-in-deal-to-acquire-calgene.html (accessed 1 December 2017).

Asveld, L., Ganzevles, J. and Osseweijer, P. (2015) Trustworthiness and responsible research and innovation: The case of the bio-economy. *Journal of Agricultural and Environmental Ethics* 28, 571–588. doi:10.1007/s10806-015-9542-2

Bawa, A.S. and Anilakumar, K.R. (2013) Genetically modified foods: Safety, risks and public concerns – A review. *Journal of Food Science and Technology* 50, 1035–1046. doi:10.1007/s13197-012-0899-1

BBC News (1996) First GM food goes on sale in UK. BBC News, 5 February. Available at: http://news.bbc.co.uk/onthisday/hi/dates/stories/february/5/newsid_4647000/4647390.stm (accessed 1 December 2017).

Beauchamp, T. and Childress, J. (2001) *Principles of Biomedical Ethics*. Oxford University Press, New York.

Beekman, V. and Brom, F.W.A. (2007) Ethical tools to support systematic public deliberations about the ethical aspects of agricultural biotechnologies. *Journal of Agricultural and Environmental Ethics* 20, 3–12. doi:10.1007/s10806-006-9024-7

Belch, G. and Belch, M. (2012) *Advertising and Promotion: An Integrated Marketing Communications Perspective*, 11th edn. McGraw Hill, New York.

Bennett, A.B., Chi-Ham, C., Barrows, G., Sexton, S. and Zilberman, D. (2013) Agricultural biotechnology: Economics, environment, ethics, and the future. *Annual Review of Environment and*

Resources 38, 249–281. doi:10.1146/annurev-environ-050912-124612

Berlin, J. (2015) Can this scientist unite genetic engineers and organic farmers? *National Geographic*, 4 May. Available at: https://news.nationalgeographic.com/2015/05/150502-nginnovators-rice-genetic-engineering-gm-organic-farming-pamela-ronald/?_ga=2.61242540.167065903. 1511123916-1086146769.1511123916(accessed 1 December 2017).

Blancke, S. *et al.* (2015) Fatal attraction: the intuitive appeal of GMO opposition. *Trends in Plant Science* 20, 414–418. doi:10.1016/j.tplants.2015.03.011

Bradley, K. (2017) *A final report on Dicamba-injured soybean acres*. Division of Plant Sciences, University of Missouri. Available at: https://ipm.missouri.edu/IPCM/2017/10/final_report_dicamba_injured_soybean/ (accessed 1 December 2017).

Bruening, G. and Lyons, J.M. (2000) The case of the FLVR SAVR tomato. *California Agriculture* 4, 6–7. http://calag.ucanr.edu/Archive/?article=ca.v054n04p6

Buzalka, M. (2000) Does GE bring good things to life? *Food Management* 35, 10. Available at: https://search.proquest.com/docview/215888687?accountid=14576 (accessed 1 December 2017).

Carter, C.A. and Gruère, G.P. (2003) Mandatory labeling of genetically modified foods: Does it really provide consumer choice? *AgBioForum* 6, 68–70. http://www.agbioforum.org/v6n12/v6n12a13-carter.pdf

Center for Environmental Risk Assessment (CERA) (2015) *Solanum lycopersicum* Tomato. GM Crop Database. Available at: http://cera-gmc.org/GmCropDatabaseEvent/FLAVR%20SAVR (accessed 1 December 2017).

Center for Food Safety (2005) *Monsanto vs. U.S. Farmers*. CFS, Washington, DC. Available at: https://www.centerforfoodsafety.org/files/cfsmonsantovsfarmerreport11305.pdf (accessed 1 December 2017).

Costa-Font, M., Gil, J.M. and Trail, W.B. (2008) Consumer acceptance, valuation of and attitudes towards genetically modified food: Review and implications for food policy. *Food Policy* 33, 99–111. doi:10.1016/j.foodpol.2007.07.002

Costanigro, M. and Lusk, J.L. (2014) The signaling effect of mandatory labels on genetically engineered food. *Journal of Food Policy* 49, 259–267. doi:10.1016/j.foodpol.2014.08.005

Dewey, C. (2017) The government is going to counter 'misinformation' about GMO foods. *Washington Post*, 3 May. Available at: https://www.washingtonpost.com/news/wonk/wp/2017/05/03/the-government-is-going-to-try-to-convince-you-to-like-gmo-foods/?utm_term=.09a7072ff38d (accessed 1 December 2017).

Dieterle, J.M. (2016) Autonomy, values, and food choice. *Journal of Agricultural and Environmental Ethics* 29, 349–367. doi:10.1007/s10806-016-9610-2

Dryzek, J.S. *et al.* (2009) Promethean elites encounter precautionary publics: The case of GM foods. *Science, Technology & Human Values* 34, 263–288. doi:10.1177/0162243907310297

Dubock, A. (2014) The politics of Golden Rice. *GM Crops & Food* 5, 210–222. http://www.goldenrice.org/PDFs/Dubock-Politics_of_GR-2014.pdf

Elderidge, S. (2003) *Food Biotechnology: Current Issues and Perspectives*. Nova Science Publishers, New York.

Ewen, S.W.B. and Pusztai, A. (1999) Effect of diets containing genetically modified potatoes expressing *Galanthus nivalis* lectin on rat small intestine. *Lancet* 354, 1353–1354. doi:10.1016/S0140-6736(98)05860-7

FAO (2017) The state of food security and nutrition in the world 2017. Building resilience for peace and food security. Available at: https://www.unicef.org/publications/files/State_of_Food_Security_and_Nutrition_in_the_World_2017.pdf (accessed 1 December 2017).

FDA (2009) POSILAC: For increasing production of marketable milk in lactating dairy cows. Available at: https://www.fda.gov/downloads/Animal Veterinary/Products/ApprovedAnimalDrug Products/FOIADrugSummaries/UCM050022.pdf (accessed 1 December 2017).

FDA (2015a) Guidance for industry: voluntary labeling indicating whether foods have or have not been derived from genetically engineered plants. Available at: https://www.fda.gov/Food/Guidance Regulation/GuidanceDocumentsRegulatory Information/LabelingNutrition/ucm059098.htm (accessed 1 December 2017).

FDA (2015b) Draft guidance for industry: voluntary labeling indicating whether food has or has not been derived from genetically engineered Atlantic salmon. Available at: https://www.fda.gov/Food/GuidanceRegulation/GuidanceDocuments RegulatoryInformation/ucm469802.htm (accessed 1 December 2017).

Fedoroff, N.V. (2015) Food in a future of 10 billion. *Agriculture & Food Security* 4, 11. doi:10.1186/s40066-015-0031-7

Fernandez-Cornejo, J. *et al.* (2014) *Genetically engineered crops in the United States*. USDA ERS, Washington, DC. http://www.ers.usda.gov/publications/pub-details/?pubid=45182 (accessed 5 April 2018).

Fischer, K., Ekener-Petersen, E., Rydhmer, L. and Björnberg, K.E. (2015) Social impacts of GM crops in agriculture: A systematic literature review. *Sustainability* 7, 8598–8620. doi:10.3390/su7078598

Flitter, E. (2017) Special Reuters report: The path to Monsanto's weed-killer crisis. *The Insurance Journal*, 9 November. Available at: https://www.insurancejournal.com/news/national/2017/11/09/470807.htm (accessed 1 December 2017).

Goddard, A. (1998) A puree genius at work. *Times Higher Education*, 17 July. Available at: https://www.timeshighereducation.com/features/a-puree-genius-at-his-work/108313.article (accessed 1 December 2017).

Grove-White, R. and Szerszynski, G.W. (1992) Getting behind environmental ethics. *Environmental Values* 1, 285–296. doi:10.3197/096327192776680016

Gupta, N., Fischer, A.R.H. and Frewer, L.J. (2011) Socio-psychological determinants of public acceptance of technologies: A review. *Public Understanding of Science* 21, 782–795. doi:10.1177/0963662510392485

Hails, R. and Kinderlerer, J. (2003) The GM public debate: Context and communication strategies. *Nature Reviews Genetics* 4, 819. doi:10.1038/nrg1182

Hakim, D. (2017) Monsanto's weed killer, Dicamba, divides farmers. *New York Times*, 21 September. Available at: https://www.nytimes.com/2017/09/21/business/monsanto-dicamba-weed-killer.html?_r=0 (accessed 1 December 2017).

Hansen, J. *et al.* (2003) Beyond the knowledge deficit: Recent research into lay and expert attitudes to food risks. *Appetite* 41, 111–121. doi:10.1016/S0195-6663(03)00079-5

Hepting, D.H., Jaffe, J.A. and Maciag, T. (2014) Operationalizing ethics in food choice decisions. *Journal of Agricultural and Environmental Ethics* 27, 453–469. doi:10.1007/s10806-013-9473-8

Heydens, W. (n.d.) Monsanto Papers (email). Available at: http://baumhedlundlaw.com/pdf/monsanto-documents/22-Internal-Email-Demonstrating-Monsanto-Ghostwriting-Article-Criticizing-IARC-for-Press.pdf (accessed 1 December 2017).

Hilbeck, A. *et al.* (2015) No scientific consensus on GMO safety. *Environmental Sciences Europe* 27, 1–6. doi:10.1186/s12302-014-0034-1

Jacobsen, S., Sørensen, M., Pedersen, S.M. and Weiner, J. (2013) Feeding the world: genetically modified crops versus agricultural biodiversity. *Agronomy for Sustainable Development* 33, 651–662. doi:10.1007/s13593-013-0138-9

Jenkins, D. (n.d.) Monsanto Papers (email). Available at: https://www.baumhedlundlaw.com/toxic-tort-law/monsanto-roundup-lawsuit/monsanto-secret-documents-page-ten/ (accessed 1 December 2017).

Just Label It (n.d.) Labeling around the world. Available at: http://www.justlabelit.org/right-to-know-center/labeling-around-the-world/ (accessed 22 November 2017).

Klümper, W. and Qaim, M. (2014) A meta-analysis of the impacts of genetically modified crops. *PLoS One* 9. doi:10.1371/journal.pone.0111629

Kolodinsky, J. (2007) Biotechnology and consumer information. In: Brossard, D., Shanahan, J. and Nesbitt, T.C. (eds) *The Public, the Media and Agricultural Biotechnology*. CAB International, Wallingford, UK, pp. 161–178.

Kolodinsky, J. (2008) Affect or information? Labeling policy and consumer valuation of rBST free and organic characteristics of milk. *Food Policy* 33, 616–623. doi:10.1016/j.foodpol.2008.07.002

Kolodinsky, J. (2015) Study: GM food labels do not act as a warning to consumers. *The Conversation*, 29 July. Available at: https://theconversation.com/study-gm-food-labels-do-not-act-as-a-warning-to-consumers-45283 (accessed 1 December 2017).

Kolodinsky, J. (2017) How did mandatory labeling impact support/opposition for the use of GE in food? Presentation at the Agricultural and Applied Economics meetings, July 2017, Chicago, Illinois.

Kolodinsky, J., Wang, Q. and Conner, D. (1998) rBST labeling and notification: Lessons from Vermont. *Choices* 13, 38–40. http://www.jstor.org/stable/43663293

Krimsky, S. (2015) An illusory consensus behind GMO health assessment. *Science, Technology, & Human Values* 40, 883–914. doi:10.1177/0162243915598381

Lambrecht, B. (2001) Outgoing secretary says agency's top issue is genetically modified food. *St. Louis Post-Dispatch*, 25 January. Available at: https://www.iatp.org/news/outgoing-secretary-says-agencys-top-issue-is-genetically-modified-food (accessed 1 December 2017).

Landrum, A.R. and Hallman, W.K. (2017) Engaging in effective science communication: A response to Blancke et al. on deproblematizing GMOs. *Trends in Biotechnology* 35, 378–379. doi:10.1016/lj.tibtech.2017.01.006

Lang, J.T. and Hallman, W.K. (2005) Who does the public trust? The case of genetically modified food in the United States. *Risk Analysis* 25, 1241–1252. doi:10.1111/j.1539-6924.2005.00668.x

Leary, W. (1994) F.D.A. approves altered tomato that will remain fresh longer. *The New York Times*, 18 May. Available at: http://www.nytimes.com/1994/05/19/us/fda-approves-altered-tomato-that-will-remain-fresh-longer.html (accessed 1 December 2017).

Levidow, L. and Carr, S. (1997) How biotechnology regulation sets a risk/ethics boundary. *Agriculture and Human Values* 14, 29–43. doi:10.1023/A:1007394812312

Liaukonyte, J., Streletskaya, N.A. and Kaiser, H.M. (2015) Noisy information signals and endogenous preferences for labeled attributes. *Journal of Agricultural and Resource Economics* 40, 179–202.

Lusk, J.L. and Briggeman, B.C. (2009) Food values. *American Journal of Agricultural Economics* 91, 184–196. doi:10.1111/j.1467-8276.2008.01175.x

Marris, C. (2001) Public views on GMOs: Deconstructing the myths. *EMBO Reports* 2, 545–548. doi:10.1093/embo-reports/kve142

Medlock, J., Downey, R. and Einsiedel, E. (2007) Governing controversial technologies: consensus conferences as a communications tool. In: Brossard, D., Shanahan, J. and Nesbitt, T. (eds) *The Public, the Media and Agricultural Biotechnology*. CAB International, Wallingford, UK, pp. 308–326.

Mepham, B. (2000) A framework for the ethical analysis of novel foods: The ethical matrix. *Journal of Agricultural and Environmental Ethics* 12, 165–176. doi:10.1023/A:1009542714497

Mericle, J. (2017) Iowa state extension forester warns of Dicamba dangers. *The Hawk Eye*, 9 October. Available at: http://www.thehawkeye.com/news/20171009/iowa-state-extension-forester-warns-of-dicamba-dangers (accessed 1 December 2017).

Modin, P.G. and Hansson, S.O. (2011) Moral and instrumental norms in food risk communication. *Journal of Business Ethics* 101, 313–324. doi:10.1007/s10551-010-0724-6

Monsanto (n.d.) What caused the failure of the Flavr Savr tomato back in the 90's and are there plans to revive it back? Available at: https://monsanto.com/innovations/research-development/q/what-caused-the-failure-of-the-flavr-savr-tomato-back-in-the-90s-and-are-there-plans-to-revive-it-back/ (accessed 24 November 2017).

Nash, M. (2000) Grains of hope. *Time*, 23 July. Available at: http://content.time.com/time/magazine/article/0,9171,50576,00.html (accessed 1 December 2017).

National Academies of Sciences, Engineering, and Medicine (2016) *Genetically Engineered Crops: Experiences and Prospects*. NAS, Washington, DC. doi:10.17226/23395

Non-GMO Project (n.d.) Verification FAQs. Available at: https://www.nongmoproject.org/product-verification/verification-faqs/ (accessed 22 November 2017).

Nosowitz, D. (2017) Evidence mounts that Monsanto's Dicamba is killing trees, too. *Modern Farmer*, 13 October. Available at: https://modernfarmer.com/2017/10/evidence-mounts-monsantos-dicamba-killing-trees/ (accessed 1 December 2017).

Robbins, J. (2011) Can GMOs help end world hunger? *Huffington Post*, 1 October. Available at: https://www.huffingtonpost.com/john-robbins/gmo-food_b_914968.html (accessed 1 December 2017).

Rollin, B.E. (2014) The perfect storm – genetic engineering, science, and ethics. *Science & Education* 23, 509–517. doi:10.1007/s11191-012-9511-3

Science and Technology Committee (1999) Science and Technology – First Report. UK Parliament House of Commons, London. Available at: https://publications.parliament.uk/pa/cm199899/cmselect/cmsctech/286/28602.htm (accessed 1 December 2017).

Slovic, P. (1987) Perception of risk. *Science* 236, 236–285. doi:10.1126/science.3563507

Stone, G.D. and Glover, D. (2017) Disembedding grain: Golden Rice, the Green Revolution, and heirloom seeds in the Philippines. *Agriculture and Human Values* 34, 87–102. doi:10.1007/s10460-016-9696-1

Taleb, N. (2010) *The Black Swan*, 2nd edn. Random House, New York.

Thompson, P.B. (1997) Food biotechnology's challenge to cultural integrity and individual consent. *Hastings Center Report* 27, 34–39. doi:10.2307/3528777

Tversky, A. and Kahneman, D. (1974) Judgment under uncertainty: Heuristics and biases. *Science* 185(4157), 1124–1131. doi:10.1126/science.185.4157.1124

United Nations (UN) (1992) Report of the United Nations conference on environment and development. Available at: http://www.un.org/documents/ga/conf151/aconf15126-1annex1.htm (accessed 1 December 2017).

United Nations (UN) (2017) The sustainable development goals report. Available at: https://unstats.un.org/sdgs/files/report/2017/TheSustainableDevelopmentGoalsReport2017.pdf (accessed 1 December 2017)

United States Department of Agriculture (USDA) (n.d.) GMO disclosure & labeling. Available at: https://www.ams.usda.gov/rules-regulations/gmo (accessed 22 November 2017).

Vermont General Assembly (2014) H.112 (Act 120): An act relating to the labeling of food produced with genetic engineering. Available at: http://www.leg.state.vt.us/docs/2014/Acts/ACT120.pdf (accessed 1 December 2017).

Waldman, P., Stecker, T. and Rosenblatt, J. (2017) Monsanto was its own ghostwriter for some safety reviews. *Bloomberg Business*

Week, 9 August. Available at: https://www. bloomberg.com/news/articles/2017-08-09/ monsanto-was-its-own-ghostwriter-for-some-safety-reviews (accessed 1 December 2017).

Ward, E.M. (2018) Glyphosate use and cancer incidence in the agricultural health study: An epidemiologic perspective. *JNCI: Journal of the National Cancer Institute* 110, djx247. doi:10.1093/jnci/djx247

Williams, G.M., Kroes, R. and Munro, I.C. (2000) Safety evaluation and risk assessment of the herbicide Roundup 1 and its active ingredient, Glyphosate, for humans. *Regulatory Toxicology and Pharmacology* 31, 117–165. doi:10.1006/rtph.1999.1371

Williams, G.M. *et al.* (2016) A review of the carcinogenic potential of glyphosate by four independent expert panels and comparison to the IARC assessment. *Critical Reviews in Toxicology* 46, 3–20. doi:10.1080/10408444.2016.1214677

Wohl, J.B. (1998) Consumers' decision-making and risk perceptions regarding foods produced with biotechnology. *Journal of Consumer Policy* 21, 387–404. doi:10.1023/A:1006904622571

2 Against the (GM) Grain: Ethical Tensions and Agrobiotechnology Activism in the USA

Bradley Martin Jones*

Department of Anthropology, Washington University in St Louis, St Louis, Missouri, USA

Introduction

On Earth Day 2012, several dozen East Bay activists smashed through the gate of University of California (UC) Berkeley's Gill Tract, established a makeshift camp and began planting organic vegetables. Rallying behind such slogans as 'We dig the farm' and 'Whose farm, our farm', the guerilla gardeners sought to challenge both the continued corporate enclosure of commons and what they perceived as a nefarious alliance between profit-driven industry and a public land-grant university. At stake in the conflict was not only the future of this particular ten-acre property but also the role of engaged citizenry in deciding how public land should be put to use. The privileging of patent-oriented biotech research over more 'natural', more egalitarian agricultural experimentation was a critical flashpoint of the activists' critique.

The following spring, hundreds of thousands of concerned citizens in cities across the world took to the streets to 'March Against Monsanto'. Elaborate displays of street theater ensued with children dressed as Monarch butterflies harassed by corporate henchmen. Enormous banners nearly blotted out the sun, exclaiming 'Say no to GMO' and 'Monsanto: Evil Seed of Corporate Greed'. The protests came on the heels of the so-called 'Monsanto Protection Act', a clause that banned US courts from halting the sale of genetically modified organisms and served to only fan the flames of an already incensed public concerned about the company's receipt of corporate subsidies, political favoritism and monopolistic reach. The mass rally was directed at (largely) American hearts and minds, designed to bring heretofore behind-closed-doors debates over biotechnology squarely, and stridently, into the public sphere.

Challenges to the agroindustrial food system in the USA have taken many forms in recent years. Some initiatives aim to support sustainability-minded family farms directly, to encourage underrepresented farmer populations or to promote organic farming, community-supported agriculture and locally oriented farmers' markets (Henderson, 2000). Other interventions seek to democratize access to healthy food through consumer price subsidy programs or through maintaining community gardens. Still other initiatives fund and conduct research on ecologically and economically sustainable growing practices at the land grant level or encourage permaculture and biodynamic approaches. More recently, cooperative buying clubs and cooperative farming operations have proliferated alongside product labeling regimes, know-your-farmer campaigns and 'slow' eating/living movements (Grasseni, 2013).

* E-mail: bradleyjones@wustl.edu

This chapter brings attention to a particular aspect of this broader alternative food movement by focusing on resistance to agricultural biotechnology, specifically on forms of direct action and public protest. As the following section will show, a growing body of literature has productively explored ethical tensions and social struggles over genetically modified organisms in Europe, Latin America, India and elsewhere. However, the USA, ground zero of agrobiotech development and proliferation, has received comparatively little attention. In this chapter I attend to two forms of public dissent that exemplify opposite ends of a spectrum of biotechnology resistance: Occupy The Farm and March Against Monsanto. The former represents a localized form of direct action to reclaim public land for sustainable and community-oriented food production while directly criticizing academic capitalism and the biotech research agenda. The latter depicts an annual mass protest designed to raise public awareness, garner widespread support and challenge the agricultural leviathan Monsanto and the constellation of values it stands for (and, in the activist rhetoric, been made to stand for). Both highlight that ethical tensions of new technology are becoming increasingly central to public critique.

In this chapter, I contribute to the growing body of literature that considers food to be an important catalyst through which activist subjectivities are materialized and around which activist practices are mobilized (Belasco, 2006; Counihan and Forson, 2011; Counihan and Siniscalchi, 2014). As anthropologist Richard Wilk suggests, 'Food has long been a focus for political and social movements in many parts of the world; food is a potent symbol of what ails society, a way of making abstract issues like class or exploitation into material, visceral reality' (2006, p. 21). But while grassroots activist practices of food systems critique have been a potent symbol and an important locus of political engagement 'in many parts of the world', such resistance has been a marginal form of critical public discourse in the USA. This is in part attributable to the fact that, as opposed to Europe and elsewhere, until recently there have been limited challenges to the agroindustrial food system around which solidarity might coalesce and a narrow imagination of what legitimate alternatives to it would look like in practice.

In shedding light on anti-biotechnology activist engagements, both ephemeral and ongoing, this chapter emphasizes three interrelated claims: that civil society critique of the agroindustrial food system is not only alive but arguably intensifying in the USA; that a key reason for this emerging resistance is an expanding sense of what counts as relevant for mobilizing dissent (namely ethical and other 'extra-scientific' factors); and finally, that these observations can be productively put to the service of stoking the 'radical imagination' – 'the ability to imagine the world, life and social institutions not as they are but as they might otherwise be' (Haiven and Khasnabish, 2014, p. 3). In highlighting the role of agrobiotechnologies as a flashpoint of sovereignty 'in many parts of the world' (as I do in the next section) and in attending to how activist mobilizations have recently been underway in the USA (as the case studies reveal), this chapter not only contributes to a lacuna within the growing body of literature on anti-biotechnology activism, it also emphasizes that genetically modified crops must be a primary area of focus for any study of food activism more broadly.

Agrobiotechnologies, I argue, have become a 'condensation symbol', an image or idea that engages an individual's most basic values and is endowed with intense effective and affective power (cf. Graber, 1976). Genetically modified seeds condense into a single referent a range of disparate meanings, including nature and culture, hubris and ingenuity, scarcity and abundance, Prometheanism and progress, apocalypse and salvation. They are thus uniquely charged with the capacity to invoke and provoke profound ethical tensions. The GM seed is also a 'condensation object' in that it catalyzes diffuse ethical actors into mobilization and solidarity around it. Whether framed as miracles or monsters, GM seeds trigger reactions out of latent states much as condensation is a polarized change of matter into denser forms, as water vapor solidifies onto glass. But this process is not simple chemistry operating by natural law. Rather than intrinsic to the technology itself, the position of biotechnologies as symbols and objects of ethical tension is in important ways an outcome of discursive framing by a variety of interested actors. That is, new technologies are positioned to perform important rhetorical work that either neutralizes activism or profoundly incites it. These case studies highlight that while early scientific framings of harm and risk have conditioned the limited public debate in the USA, social and ethical factors are increasingly emerging as genuine arenas of democratic dissent. Moreover, a discursive framework that limited public resistance by narrowing the field for what was and was not seen as legitimate grounds for mustering resistance is widening to include 'extra-scientific' factors, thus affording possibilities for public engagement around diverse ethical issues relevant to broader biotech deliberations.

The polarizing discourses in which biotechnologies have become deeply embedded frame the public debate in a way that results in rampant schismogenesis – the self-amplifying process of divergence (Stone, 2015). Indeed, public debate is a rather generous term for disputes frequently more militant than civil. The case studies offered here point to some of the ways in which activist practices mobilize around agrobiotechnologies in particular, peculiar and historically contingent ways. They also make it clear that ethical tensions (e.g. community food sovereignty) and cultural anxieties (e.g. ontologies of the natural world and the human place in it) are increasingly important in the political imaginary of emerging counterpublics. As Stone (2010) has persuasively argued, because genetically modified crops have had significant impacts on industry/academy relations, rural environments, farmer livelihoods and civil society, anthropological engagement with how debates get framed and forwarded in the public sphere should be of primary importance on the research agenda.

Situating the Struggle in a Global Context

Social movements seeking to resist agrobiotechnology have sprouted across the globe since the release of the first transgenic crops two decades ago. England is a well-studied example of how a confluence of socially embedded and historically contingent factors can influence popular reception of GM crops at a critical time. As Stone (2002a) highlights, a combination of the biotech industry's heavy-handed brashness, the choice to traffic in unlabeled modified crops, the limited perceived value of GM varieties to consumers and the growing networking capabilities enabled by the internet all contributed to a context in which agrobiotechnology faced significant criticism from the outset in the UK. Moreover, the timing of GM crops' appearance across the pond could hardly have been worse. Fresh on the heels of the BSE debacle, in which despite avowed public health expert testimony to the contrary it was revealed that Mad Cow disease did in fact pose a serious human health risk, the public's faith in the established regulatory system to mitigate risk was severely compromised. What's more, because BSE was linked to modern industrial agricultural practices, 'modern' and 'unnatural' became discursively linked in a manner that framed the public perception of genetically modified foods (Jasanoff, 2005, p. 122). Alongside Prince Charles' outspoken advocacy against tampering with nature (interfering with the designs of the Creator, as he phrased it) and the British government's decision to encourage a participatory, deliberative and issue-specific debate around GM foods, it became abundantly clear that the British public was concerned about far more than scientific considerations. As Jasanoff reflects, 'In Britain, where the public outcry against GM was both sharpest and most explicitly tied to the values of tradition and countryside, it seemed certain that commercial GM agriculture would henceforth have to answer public questions about need, benefit, and level of certainty that it could fulfill its promise of health and environmental safety' (2005, p. 144).

A similar concern over ethical and social factors that extend beyond a narrow scientific assessment of risk buttressed the critical resistance towards GM crops that unfolded in France. As Heller (2013) instructively shows, this framing has largely come not from consumers but at the behest of an established network of rural farmers concerned that GM crops would only intensify the deleterious effects of industrial agriculture on smallholder farmers. The Confederation Paysanne established a discourse of what Heller calls a 'rationality of solidarity' that challenged neoliberal economic models based on profit and vertical integration. Through the staging of direct actions (e.g. GM crop pulls) and public protests (e.g. peasant rallies at the foot of the symbolic Eiffel Tower), the social movement has effectively positioned food within a cultural framework, linking anti-GMO activism with alter-globalization debates. The effect has been a shift in the locus of expertise. As Heller notes, 'For the first time in the debate [initially conceived as a problem of calculable risk] paysan farmers, as well as scientists, could speak with cultural authority about GMOs' (2013, p. 32). This broadening of expertise democratized and catalyzed public critique, and stiffened the opposition to genetically modified foods which were framed rhetorically as injurious to the peasant farmer, who was increasingly positioned as a harbinger of national values.

The role of producers in contesting GM technologies is also prominent in Latin America, with Brazil and Argentina being two of the three largest producers of genetically modified crops worldwide. As in the UK and France, here too we see the broadening of significant factors when it comes to judging the merits of agrobiotechnology. In Mexico, for instance, the debate has largely been framed in terms of 'biopollution' (Kinchy, 2012). As a nation with a robust smallholder and subsistence farming population and an unparalleled social, economic and ethical investment in landrace grain varieties (particularly maize), the Mexican public has expressed vociferous criticism of GM seeds. Of utmost concern is the risk of transgene flow and genetic contamination, a hazard made all the more relevant by the controversial Quist and Chapela (2001) finding that transgenes were present in native varietals even several years after a nationwide moratorium. Fitting (2011) has shown that the broader ethical and economic context of maize biodiversity in the Mexican imaginary engendered widespread rural resistance to GM corn and situated its introduction as an affront to local food security, national economic sovereignty and cultural survival. Like the alter-globalization critique expressed in France, Mexican farmers rallied under a banner critical of global agricultural restructuring, trade liberalization, foreign control of seeds and the penetration of capital intensive agriculture. As 'biopollution' has framed the debate in Mexico, in Colombia 'biopiracy' came to shape much of the public discourse on agrobiotechnologies. Gene and germplasm prospecting from researchers and corporations in the Global North has entrenched the defensive stance of Colombian resistance movements against myriad forms of neoliberal economic advances, including the introduction of GM seeds (Heller and Escobar, 2003). Local forms of identity politics animated by global development initiatives have foregrounded cultural difference, local knowledge, self-determination and conservation. The result, much as in Mexico, France and elsewhere, is the framing of the agrobiotechnology in cultural, social, ethical and ecological terms that have buttressed resistance to GM crops.

Turning briefly from food to fiber production, biotechnology resistance in India has done much to shape the discourse of the global debate. This is not only as a result of prominent activist ideologues like Vandana Shiva, whose vociferous critiques are regularly echoed and amplified by GMO opponents worldwide, but also because India has been put forward as symbolic of the potential of GMOs in the Global South by biotech companies looking to reframe the international narrative and reclaim their public image (Stone, 2002a). In the hands of Shiva and her legion, India has become as much a test plot for staging counternarratives to corporate dominant discourse by emphasizing the deleterious effects of GMO adoption on peasant farmer livelihoods. Billed as encouraging 'bioserfdom' or 'biofeudalism', Bt cotton became characterized as 'suicide seed', not least because of rampant farmer suicides in the Warangal District of Andhra Pradesh that set the stage for the introduction of GMOs (Herring, 2015; Stone, 2002b). Operation Cremate Monsanto, a national farmer protest movement initiated in fall of 1998, led to a number of direct actions, including the burning of test crops and a series of public trials (Herring, 2006). Though ultimately unsuccessful in prohibiting the introduction of transgenic cotton, the protest did work to rhetorically situate agrobiotechnology as a threat to national independence, smallholder farmers, local ecologies and human health, much to the chagrin of biotech companies who were proposing precisely the opposite. In all, biotech activism in India had the effect of loading the term GMO with political and moral baggage that has profoundly impacted the state of the debate in the Global South and elsewhere (Schurman and Munro, 2010, p. 185).

This comparative overview highlights that ethical tensions and 'extra-scientific' factors have profoundly influenced the framing of the public debate, the manifestation of dissent and the resultant regulatory approaches taken towards GM foods (see also Stone, 2010). What is striking then is the way in which the debate in the USA has been staged in terms of scientific risk assessment from the outset. Arguably, much of this result can be attributed to path dependency, specifically the early framing achieved at the seminal Asimolar conference in 1975 in which the country's leading biotech researchers came together to review scientific progress and set out general principles of regulation. That scientific researchers were concerned with scientific responsibility and scientific freedom is perhaps unsurprising. But as Jasanoff (2005) has argued, what is remarkable is that an event with such restricted representation provided 'crucial conceptual and procedural raw material for structures that continued to operate when the world of biotechnology had grown a great deal more complex' (p. 47). As opposed to Europe and elsewhere, 'that biotechnology might one day destabilize basic elements of

social order – kinship, for example, or farmers' rights to own and sow seeds – was very far from the thoughts of the field's founding fathers' (p. 47). By the time the public took notice in the mid-1990s, she suggests 'the GM horse had largely fled the regulatory barn' (p. 144).

But not entirely. Scientific risk assessments of biological control and the repercussions of biotechnologies on human and, to a lesser extent, environmental health certainly did dictate national policies. They also framed the dominant discourse to the extent that certain forms of public dissent became untenable. As a result, in the USA, as is so often the case under the neoliberal regime, public perception was vetted through the market. The tremendous growth of the organic sector in the years of GMO roll-out led to the development of the National Organic Program and ostensibly provided a space for popular democracy to influence agricultural production and food manufacture. While genetically engineered products would remain unlabeled, their ideological antithesis would be branded in bold. And while 'substantial equivalence' would come to define the national regulatory approach to these novel forms of life, 'organic' would reflect an ontological assertion held by the anti-biotech community that non-GM products were a good deal more 'pure' or 'natural'. In the end, the ethical and extra-scientific values that had produced so much friction on the international stage were in the USA channeled into a form – commodities and market behavior – perfectly accommodating to capital. And as numerous academic and public activists have persuasively argued, the growth (and subsequent cooptation) of the organic sector works to reproduce the neoliberal subjectivities and economic forms the organic ideal is ostensibly designed to contest (Guthman, 2008; Obach, 2015).

More than simply laying a fertile foundation for organics, neoliberalism shaped the US regulatory and resistance environment of biotechnology in profound ways. Indeed, core neoliberal tenants such as strong patent protections (in this case, on life itself), novel academic–industry relations and a speculative growth economy all provide the context in which biotechnologies came on the scene, and their logics contributed to the conditions for which GMOs would not only stay but thrive there. Directly influencing the state of struggle, as Schurman and Munro (2010) suggest, the limited activism that did crop up in the USA was enormously unsuccessful in slowing or reversing the proliferation of GM crops. Though Schurman and Munro wish to argue that anti-biotech sentiments did influence agrobiotechnologies in subtle and diffuse ways, they nevertheless emphasize that when 'looked at from the vantage point of 2008, little evidence indicates that [the US anti-biotech movement] had much of an influence at all' (2010, p. 118).

I wish to suggest that by 2012–13 the scene had changed, potentially dramatically. The two case studies presented here indicate that anti-biotech activism is not only alive in the USA but arguably advancing. Buttressed by novel digital technologies such as social media, impacted by the global economic recession of the late 2000s and influenced by both short-term (yet resonant) social formations like the Occupy Movement as well as more protracted sensibilities born by the arrival of the Anthropocene, Occupy The Farm and the March Against Monsanto are caught up in the turbulence of a perfect storm for political protest. Despite the biotech industry's active attempt to frame issues in purely scientific terms (Kinchy, 2012), might ethical and 'extra-scientific' factors finally be creeping into public debate? Might the core tenets of neoliberalism that buttressed the rise of biotechnologies also, in an era of arguably shifting political winds, lay the foundation for their demise? Might novel protest practices oriented towards alterity and agency expand the otherwise beleaguered horizon of political possibility? These are questions certainly up for critical debate and on which Occupy The Farm and March Against Monsanto may shed some revelatory light.

Occupy The Farm

Occupy The Farm was a three-week 'occupation' of the UC Berkeley-owned Gill Tract that commenced on Earth Day (22 April) 2012. The action involved several dozen people who established a provisional camp and planted some 15,000 vegetable starts, with the intention of simultaneously expressing discontent over years of failed negotiation on how best to use the land and acting directly to bring the reality the activists envisioned into being. As one occupier described in an interview for the 2014 documentary on the protest, 'The whole point of this is to not talk about what we want, to not demand what we want, but to make what we want real' (Occupy The Farm, 2014). In this way, Occupy The Farm was conceived as a direct action. As anthropologist David Graeber defines it, direct action 'is the insistence, when faced with structures of unjust authority, on acting as if one is already free. One does not solicit the state. One does not even necessarily make a grand gesture of defiance. Insofar as one is capable, one proceeds as if the state does not exist' (2009, p. 203). It was also oriented to a prefigurative politics, the idea that means must be consistent with ends and that 'the organizational form that an activist group takes should embody the kind of society [they] wish to create' (Graeber, 2013, p. 23). Occupy The Farm manifested as a radical form of political protest in which traditional reform channels were circumvented for direct action.

The radical form the Occupy The Farm occupation eventually took was not for a lack of effort to proceed more conventionally. Indeed, Occupy The Farm was born of gradual desperation conditioned by a decade and a half of foiled attempts to convert the plot into a living laboratory for sustainable and urban farming. Resistance catalyzed in 2012 only when the UC Berkeley made its intentions clear that it would move forward with plans to develop the Tract into grocery retail, a youth sports complex and gentrified housing (SEAL, no date). Not only would this have irredeemably taken the land out of cultivation, preventing any intentions of community-based agricultural education, it would have also (according to activists) upended the university's institutional charter as a public land-grant institution and the terms of the Gill Tract endowed to the state under the auspices of its continued agricultural use. On these ethical grounds of civil disobedience, Occupy The Farm took to radical measures to stake a political claim to the ten remaining acres of fertile land in the heart of urban Albany.

Establishing the specific context of the occupation still hardly breaches the origin of popular discontent, long simmering beneath the surface and erupting under various guises. UC Berkeley has historically been a cauldron of progressive activism going back at least a half century. In 1964 Mario Savio famously declared that 'there comes a time when the operation of the machine becomes so odious, makes you so sick at heart, that you can't take part, you can't even passively take part, and you've got to put your bodies upon the gears and upon wheels, upon the levers, upon all the apparatus, and you've got to make it stop' (1964, n.p.). Direct action against liberal capitalist forms is deeply embedded in the political culture of the greater Berkeley community. It manifests here when local activists throw their bodies upon the gears of the agroindustrial machine resolute to not 'even passively take part'.

More directly germane to the Occupy The Farm action has been the three-decade dismantling of public education in California. With neoliberal rollbacks, corporate funders have increasingly coopted the land-grant's mission 'to teach such branches of learning as are related to agriculture and the mechanic arts'. Industry funding for university research has grown sharply since the passing of Bayh-Dole Act in 1980, a seminal piece of legislation that greenlighted the use of public funds to conduct research leading to private patents. UC Berkeley has been situated at the heart of the resulting changes in biomedical and biotechnological research. The Novartis deal struck in 1998, for instance, allocated tens of millions of dollars of private funding to be used for biotechnology research in exchange for dissertation review rights, licensing access to a substantial portion of department discoveries and significant representation on the research advisory committee (Rudy et al., 2007). What's more, the denial of tenure to Ignacio Chapela, who had recently published unflattering findings about the possibility of GM contamination

in landrace cultivars, was seen by many as a clear indication of UC Berkeley's increasing status as a 'kept university' (Press and Washburn, 2000). The dismantling of the Center for Biological Control on the Gill Tract itself – which for several decades had conducted pioneering research on ecologically friendly pest management systems and generally raised concerns over the pesticide treadmill endemic to industrial agriculture – was further evidence of the University's 'kept' status. Since the late 1990s, the research portion of the Tract has been used primarily for experiments in GM biotechnologies, and specifically corn genetics.

None of this was lost on the Occupy The Farm activists; quite the opposite in fact. Changes in industry–academy relationships and shifting research priorities were at the heart of occupiers' ethical critique and political discontent. As one activist complained, 'When you end up in that situation, obviously the research gets steered away from certain things that could simply be public goods towards things that are obviously going to have an impact on the bottom line of powerful corporations' (Occupy The Farm, 2014). Prominent agroecologist and Occupy The Farm supporter Miguel Altieri makes the point even more forcefully: 'That corn they say is for basic research, but it's all precursory for biofuel and transgenics, I'm not going to be fooled that this is just basic research, innocent research, and that they're just trying to understand how corn works, no, no.' Deploying the ethically fraught language of capital's 'obviously' nefarious aims and of practitioner's 'innocence' or guilt, Occupy The Farm expressed critique of the university's increasing privatization and championed the idea that public land should be used to serve a public purpose – indeed, the public 'good'. Occupiers rallied around the notion that 'farmland is for farming' (Roman-Alcalá, 2015).

As pronounced on the website, the vision of the Occupy The Farm movement is multifold:

a future in which communities make use of all available land – occupying it when necessary – to create urban agriculture alternatives and meet local needs in the face of economic and environmental crisis. Our goal for the Gill Tract is to establish a productive farm with a Center for Urban Agriculture and Food Justice. We aim to preserve this rich natural resource in perpetuity, emphasizing much needed research into sustainable urban agriculture, open access, and participation by the larger East Bay community. (Occupy The Farm, no date)

By squatting on public land, occupiers hoped to bring that egalitarian and ethical vision into reality without waiting for prior consent. And shockingly, after much contention that included riot police, water blockage and several arrests, the occupation was partially successful in achieving just over an acre plot set aside for 'community-based projects'. Occupiers have since used the space, alongside other community coalitions, to provide food for neighbors in need as well as to conduct research on the efficacy of sustainable agroecological approaches. As of 2017, the activists continue to wage an anti-development and anti-biotechnology campaign and are committed to using direct action tactics to achieve community goals in future. What's more, their community and media outreach, including the 2014 documentary, has brought widespread attention to not only the occupation itself but to the nested ethical tensions around corporate control of seeds and science that continues to buttress the activists' resistance and shape the reality of biotech research and proliferation under the neoliberal regime.

March Against Monsanto

While Occupy The Farm was envisioned as a localized action to achieve particular community ends of food sovereignty and issue visibility, March Against Monsanto is a now annual public protest taking place in city streets the world over. The first iteration in 2013 brought out some two million concerned citizens in Chicago, Berlin, Buenos Aires and elsewhere. By May 2015 the movement had grown to rally an estimated four million protestors in over 400 cities worldwide.

Though the demonstration is global in scope, it was initiated in the USA in partial response to the so-called Monsanto Protection Act signed by President Obama a few months before the first mass movement. March Against Monsanto's stated goals are to challenge the corporate hegemony of the seed market, to contest the revolving door of industry insiders in Washington and to raise awareness of the harmful effects of GM agriculture on the health of people and the planet (Derricks, 2014).

As the self-proclaimed world leader in biotech crops, with seeds containing Monsanto traits accounting for more than 90% of the acres planted worldwide with insect-resistance and herbicide tolerance, the corporation has long been the easy target of anti-biotech animosity. Concerns over the scale of market penetration are coupled with the company's history as a prominent chemical manufacturer, leading to apprehension that biotech crops are developed with (to many critics' minds) the intent of soaking up greater quantities of the company's prominent herbicide glyphosate. Monsanto's famed bravado in entering new markets, likened to a bull in a china shop, has done nothing to ameliorate widespread hostilities (Charles, 2001). For these and other reasons, Monsanto has become a notorious public punching bag for much anti-biotech dissent.

Though Monsanto is a company activists love to hate, March Against Monsanto is the first mass action framed directly against the corporation to be conducted at this scale and certainly the most prominent in the USA to date (the more broadly conceived Seattle WTO protests notwithstanding). Significantly, the protest began from a simple act of posting a page on social media. Tami Canal, a Utah-based stay-at-home mother, disgruntled over congressional denial to afford states the right to label foods containing GM products, announced a plan to publically boycott Monsanto later that spring. She recalls, 'If I had gotten 3,000 people to join me, I would have considered that a success' (Associated Press, 2013, n.p.). But word quickly spread with the participation of other, more established anti-biotech advocacy organizations as well as anarchist groups such as Anonymous. A few short years after Tami's initial posting on Facebook, the March Against Monsanto page now has over one million followers. Social media organizing strategies have been at the core of the protest's success. Nevertheless, old-fashioned tactics have contributed to the movement's force as well, with public protest and media attention being key strategies of political persuasion. The incorporation of new technologies to achieve old ends has allowed March Against Monsanto to gain global traction in a remarkably short time span.

Discussing the broader ideological and political vision of the movement, Tami hopes that the boycott will eventually lead to an outright ban on GM foods. But, as for the consumer boycott, she insists that the market is the best way to challenge the corporation:

> People should let their friends and family know, and we need to boycott the companies (that use GMOs) to hit them where it hurts, and that's their bottom line. As a collective society we're going to say we're not going to buy that crap and not going to support companies putting irresponsible stuff out there and poisoning our kids and our bodies; we have no choice but to boycott … We have to put our money where our mouth is. (Meyer, 2013, n.p.)

Elsewhere she continues to couple the rhetoric of consumer freedom with that of democratic rights and American values, suggesting that 'Minimally I do think we need to have labelling, the consumer has the right to know what they're buying. Ultimately I really believe we need a ban in this country. Like you said, the land of the free, and here we are poisoning our hard working citizens' (The Rundown Live, 2013, n.p.). The discourse of citizen welfare and core democratic freedoms works to situate the movement within the broader public imaginary as not being simply *against* a particular company but *for* a society with broad public participation and biotech literacy. While the use of sound-bite science and affective appeals to mobilize dissent may in fact produce 'illiteracy' – in the end fanning little more

than reactionary, emotionally laden flames – the technique is consistent with the stock rhetorical approach of much grassroots activism in appealing to the hearts and minds of general publics. The rallying cry that Monsanto is guilty of 'crimes against humanity', a common claim at March Against Monsanto demonstrations, is the (il)logical conclusion of this global war of rhetoric. Ethical tensions indeed.

While the resistance rhetoric works epistemically to bring broad public attention to key movement issues centered on risk and harm (e.g. bringing attention to Seralina's controversial cancer findings or Chapela's evidence of genetic flow), it also works ideologically to buttress continued activist engagement. In an interview, Tami hits on a key point of the movement's efficacy, noting 'I believe MAM [March Against Monsanto] leaves people with tangible hope in what seems like an impossible situation at times. The world is standing up to biotech and together we will take back our right to clean food and a sustainable planet' (Rensink, 2014, n.p.). The emphasis on hope in the midst of corporate hegemony challenges not only biotechnologies but the cynicism towards political possibility that has come to characterize the neoliberal era (Rethmann, 2013). What is more, by emphasizing the collective resistance mustering the world over, the activists come to consciousness as part of a broader community. Simultaneously emphasizing possibility alongside solidarity, both of which are reinforced in the act of actual collective public protest, sustains activist subjectivities as they learn to identify with the political economic ends the movement seeks to engender. In this way, anti-biotech politics is also an identity politics where new communities of collaboration and critique are forged in the crucible of collective action (Hynes and Sharpe, 2009; Roelvink, 2010).

Conclusion

The examples of Occupy The Farm and March Against Monsanto highlighted here suggest that public-political mobilization around the issue of biotechnology in the USA is active and arguably advancing. Although much of the resistance rhetoric still frequently centers on the logic of harm, these cases make clear that ethical tensions and other 'extra-scientific' principles are increasingly on the public radar. In the case of Occupy The Farm, community food sovereignty and neoliberal federal educational policies – specifically in respect to biotech research oriented to privatization rather than the broader public good – animate activist engagement. The March Against Monsanto protest, on the other hand, criticizes political corruption for establishing corporate forms of welfare and challenges one biotech company's increasing monopoly control of seeds. While early framings of harm and risk have historically set the terms of public debate, social and ethical factors are increasingly coming to the fore as legitimate arenas of democratic dissent. In turn, a discursive framework that worked to limit public resistance by conditioning what was and was not seen as genuine grounds for mustering opposition is opening to new possibilities for public engagement around diverse ethical issues relevant to the broader biotech debate.

The shift away from a thoroughly scientific, risk-based framing in the otherwise limited US public debate is revealed as all the more necessary given arguments that the biotech industry is actively discouraging scientific risk assessment. As Stone (2014) has shown, the same alliances between academic researchers and profit-driven industries that spurred Occupy The Farm critique have agnotological effects – actively producing ignorance alongside knowledge – that lead increasingly to a state of widespread bioinsecurity. Bioinsecurity suggests 'not just that altered organisms may pose threats but that the system for studying threats has been compromised. In fact, a biotechnology establishment (comprising new types of alliances between industry and academy) is active in blocking potentially uncomfortable research on the impacts of genetic engineering' (Stone, 2014, p. 72). Without aggressive ecological and health risk research to keep pace with biotech advancements, the criteria

by which science-based critique might manifest will be increasingly non-existent. The deleterious ramifications for the public debate are clear. Clear, that is, unless critique begins to shift towards the apparatus of academic research itself and toward broader ethical tensions, a process which this chapter has suggested might in nascent form already be underway.

The activist struggles over biotechnology emphasized in this chapter should be considered as nodal points of a rhizomatic (polycentric, non-hierarchical) international grassroots social movement promoting food sovereignty (Patel, 2009; Wald, 2015). Although the seeds of this form of rights-based resistance were planted by peasant farmers in the Global South, the fight for food sovereignty is also flourishing at the margins of Empire (van der Ploeg, 2008). For instance, in Hawaii, Gupta (2015) explores place-based food sovereignty efforts that mobilize around the concept of aloha 'āina (love for the land) to protest widespread cultivation of GM seeds. Alkon and Mares (2012) highlight food sovereignty strategies within Latino communities in Seattle, Washington, and Kurtz (2015) explores the recent passing of a 'food sovereignty ordinance' as a biopolitical struggle over the nature of food systems in rural Maine. What these cases make clear is that democratic control of a socially just food system presents particular challenges for its proponents in the USA, where rights to private property are culturally entrenched, where there is a limited political bloc to voice agrarian concerns and where regulatory/bureaucratic authorities exercise strict local controls of food production. Despite these challenges, food sovereignty – with its discourses of rights and what is ethically right – increasingly animates activist engagements nationwide.

The resistance practices emphasized here also clearly resonate beyond food spheres along many of the same registers as the global Occupy movement. Though it may have receded from front-page headlines, the sentiments that animated Occupy linger – particularly the twinned crises of global capitalism and representative democracy. Like the global Occupy movement, Occupy The

Farm sought to reinvigorate democratic practices beyond established political parties and the state and to engender a prefigurative political economy that valorizes egalitarian principles. March Against Monsanto continues to bring mass public attention to issues of corporate control of what is framed as a public commons (seeds) and to global inequities created by a political economy that privileges profit-driven actors in the Global North. Both work to inspire an imagination of an alternative future and, in terms of politics and social relations, open up space for reconceptualizing the possible. Importantly, like Occupy itself, they also both rely heavily on emerging social media for aggregation, coordination and issue framing (Juris, 2012). At the same time, then, that new agricultural biotechnologies provoke ethical tensions, new digital technologies convoke the radical imagination and offer hopeful glimmers of a resolution (cf. Haiven and Khasnabish, 2014).

To this optimistic reading, skeptics may well retort 'enough with the utopian dreaming'. They might be inclined to ask not what forms resistance takes, but what tangible effects has it had on ameliorating the conditions under protest. What more just policies have been achieved by March Against Monsanto demonstrations? Has Monsanto's stock taken a hit? Has their monopoly unraveled? The same holds for the Occupy The Farm occupation. Has the 'kept university' been liberated from capital? Have the structural conditions that created community food insecurity been improved? Indeed, the truly critical scholar may well argue that public calls for labeling, market intervention and reform are perfectly in the service of the industries they are otherwise aligned to contest. Or that anarchist tactics alienate potential fence-riding allies that may otherwise be inclined to join rank. These are undoubtedly important debates to foster; critiques to which I remain sympathetic.

Yet at the risk of sounding Pollyannaish, the opposite must also be emphasized. The examples marshalled here reiterate that alternatives to neoliberal, corporate hegemony do exist and that resistance to these dominant forms has met some (however qualified)

success. Given the paucity of research on the topic of US anti-biotech resistance movements, this is an important point to stress. Challenges are caught up in a perfect storm in which new digital technologies meet novel biotechnologies, and are conditioned by global political and economic restructuring in which GM seeds have become a condensation object, a public battlefield and a hotly contested form of proxy politics. While much academic criticism of agricultural biotechnology focuses on the evolving political economy that has contributed to its rapid and increasingly global proliferation – critique that has admittedly done much to expose the (often) nefarious alliances between the academy, capitalism and political power – I would argue that a limited focus on entrenched structural violence works to make dominant forms seem unassailable. In line with the hope-oriented rhetoric offered by March Against Monsanto activists, rather than accept the neoliberal creed that 'there is no alternative' or fall prey to the often resignation-inducing effects of a 'politics of the antis' (Ferguson, 2009) – a politics that is geared towards antagonism, as for instance anti-neoliberalism or anti-globalization – activists ought also to employ a politics of the alter, the crafting of worlds otherwise. If we are inclined to consider activist engagements as contributing to a process of subject-making (a new political actor in formation) and new technologies (such as social media) as affording heretofore unimaginable mobilizations, solidarities and counterpublics, then times may be auspicious. Indeed, if diverse forms of resistance are emphasized rather than elided, and if emerging alternatives to the neoliberal status quo are recognized as legitimately powerful rather than rebuked as intrinsically impotent, then perhaps another economic, political and ethical world becomes possible.

References

Alkon, A.H. and Mares, T.M. (2012) Food sovereignty in US food movements: Radical visions and neoliberal constraints. *Agriculture and Human Values* 29, 347–359. doi:10.1007/s10460-012-9356-z

Associated Press (2013) Protestors around the world march against Monsanto. *USA Today*, 26 May. Available at: http://www.usatoday.com/story/news/world/2013/05/25/global-protests-monsanto/2361007/ (accessed 8 April 2016).

Belasco, W. (2006) *Appetite for Change: How the Counterculture Took on the Food Industry*, 2nd rev. edn. Cornell University Press, Ithaca, New York. doi:10.7591/9780801471278

Charles, D. (2001) *Lords of the Harvest: Biotech, Big Money, and the Future of Food*. Perseus, Cambridge Massachusetts.

Counihan, C. and Forson, P.W. (eds) (2011) *Taking Food Public: Redefining Foodways in a Changing World*. Routledge, New York. doi:10.4324/9781315881065

Counihan, C. and Siniscalchi, V. (eds) (2014) *Food Activism: Agency, Democracy, and Economy*. Bloomsbury, New York.

Derricks, K.L. (2014) March against Monsanto. Available at: http://www.march-against-monsanto.com/home/ (accessed 8 April 2016).

Ferguson, J. (2009) The uses of neoliberalism. *Antipode* 41, 166–184. doi:10.1111/j.1467-8330.2009.00721.x

Fitting, E. (2011) *The Struggle for Maize: Campesinos, Workers, and Transgenic Corn in the Mexican Countryside*. Duke University Press, Durham, North Carolina. doi:10.1215/9780822393863

Graber, D. (1976) *Verbal Behavior and Politics*. University of Illinois Press, Champaign, Illinois.

Graeber, D. (2009) *Direct Action: An Ethnography*. AK Press, Oakland, California.

Graeber, D. (2013) *The Democracy Project: A History, a Crisis, a Movement*. Spiegel & Grau, New York.

Grasseni, C. (2013) *Beyond Alternative Food Networks: Italy's Solidarity Purchase Groups*. Bloomsbury, New York. doi:10.5040/9781350042117

Gupta, C. (2015) Return to freedom: Anti-GMO Aloha ʻĀina activism on Molokai as an expression of place-based food sovereignty. *Globalizations* 12, 529–544. doi:10.1080/14747731.2014.957586

Guthman, J. (2008) Neoliberalism and the making of food politics in California. *Geoforum* 39, 1171–1183. doi:10.1016/j.geoforum.2006.09.002

Haiven, M. and Khasnabish, A. (2014) *The Radical Imagination: Social Movement Research in the Age of Austerity*. Zed Books, London.

Heller, C. (2013) *Food, Farms, and Solidarity: French Farmers Challenge Industrial Agriculture and Genetically Modified Crops*. Duke University Press, Durham, North Carolina. doi:10.1215/9780822394808

Heller, C. and Escobar, A. (2003) From pure genes to GMOs: Transnationalized gene landscapes in the biodiversity and transgenic food networks. In: Heath, D. and Lindee, M.S. (eds) *Genetic Nature/Culture*. University of California Press, Berkeley, California, pp. 155–175.

Henderson, E. (2000) Rebuilding local food systems from the grassroots up. In: Magdoff, F., Foster, J.B. and Buttel, F.H. (eds) *Hungry for Profit: The Agribusiness Threat to Farmers, Food, and the Environment*. Monthly Review Press, New York, pp. 175–188.

Herring, R.J. (2006) Why did 'operation cremate Monsanto' fail: Science and class in India's great terminator-technology hoax. *Critical Asian Studies* 38, 467–493. doi:10.1080/14672710601073010

Herring, R.J. (2015) Politics of biotechnology: Ideas, risk, and interest in cases from India. *AgroBioForum* 18, 142–155. http://agbioforum.org/v18n2/v18n2a02-herring.htm

Hynes, M. and Sharpe, S. (2009) Affected with joy: Evaluating the mass actions of the anti-globalization movement. *Borderlands* 8, 1–21.

Jasanoff, S. (2005) *Designs on Nature: Science and Democracy in Europe and the United States*. Princeton University Press, Princeton, New Jersey. doi:10.1515/9781400837311

Juris, J.S. (2012) Reflections on #Occupy everywhere: Social media, public space, and emerging logics of aggregation. *American Ethnologist* 39, 259–279. doi:10.1111/j.1548-1425.2012.01362.x

Kinchy, A. (2012) *Seeds, Science, and Struggle: The Global Politics of Transgenic Crops*. MIT Press, Cambridge, Massachusetts.

Kurtz, H.E. (2015) Scaling food sovereignty: Biopolitics and the struggle for local control of farm food in rural Maine. *Annals of the Association of American Geographers* 105, 859–873. doi:10.1080/00045608.2015.1022127

Meyer, N. (2013) Feature interview: March Against Monsanto founder on the future, 'pro-organic vs. anti-Monsanto' debate and more. *Althealth Works*, 2 July. Available at: http://althealthworks.com/821/feature-interview-march-against-monsanto-founder-on-the-future-pro-organic-vs-anti-monsanto-debate-and-more/ (accessed 8 April 2016).

Obach, B.K. (2015) *Organic Struggle: The Movement for Sustainable Agriculture in the United States*. MIT Press, Cambridge, MA. doi:10.7551/mitpress/9780262029094.001.0001

Occupy The Farm (2014) [documentary film]. Premier. Directed by Todd Darling.

Occupy The Farm (n.d.) Occupy The Farm. Available at: http://occupythefarm.org/ (accessed 9 April 2016).

Patel, R. (2009) Food sovereignty. *Journal of Peasant Studies* 36, 663–706. doi:10.1080/03066150903143079

Press, E. and Washburn, J. (2000) The Kept University. *The Atlantic*, March. Available at: http://www.theatlantic.com/magazine/archive/2000/03/the-kept-university/306629/ (accessed 8 April 2016).

Quist, D. and Chapela, I. (2001) Transgenic DNA introgressed into traditional maize landraces in Oaxaca, Mexico. *Nature* 414, 541–543. doi:10.1038/35107068

Rensink, E. (2014) Exclusive interview: March against Monsanto founder says Saturday to be biggest event yet. *The AntiMedia*. Available at: http://theantimedia.org/exclusive-interview-march-against-monsanto-founder-says-saturday-to-be-biggest-event-yet/ (accessed 8 April 2016).

Rethmann, P. (2013) Imagining political possibility in an age of late liberalism and cynical reason. *Reviews in Anthropology* 42, 227–242. doi:10.1080/00938157.2013.844013

Roelvink, G. (2010) Collective action and the politics of affect. *Emotion, Space, and Society* 3, 111–118. doi:10.1016/j.emospa.2009.10.004

Roman-Alcalá, A. (2015) Broadening the land question in food sovereignty to northern settings: A case study of Occupy the Farm. *Globalizations* 12, 545–558. doi:10.1080/14747731.2015.1033199

Rudy, A.P., Coppin, D., Konefal, J., Shaw, B.T., Ten Eyck, T., Harris, C. and Busch, L. (2007) *Universities in the Age of Corporate Science: The UC Berkeley-Novartis Controversy*. Temple University Press, Philadelphia, Pennsylvania.

Savio, M. (1964) On freedom and resistance. Available at: http://www.savio.org/who_was_mario.html (accessed 18 November 2017).

Schurman, R. and Munro, W.A. (2010) *Fighting for the Future of Food: Activists versus Agribusiness in the Struggle over Biotechnology*.

University of Minnesota Press, Minneapolis, Minnesota.

Stone, G.D. (2002a) Both sides now: Fallacies in the genetic modification wars, implications for developing countries, and anthropological perspectives. *Current Anthropology* 43, 611–630. doi:10.1086/341532

Stone, G.D. (2002b) Biotechnology and suicide in India. *Anthropology News* 46, 5. doi:10.1111/an.2002.43.5.5.2

Stone, G.D. (2010) The anthropology of genetically modified crops. *Annual Review of Anthropology* 39, 381–400. doi:10.1146/annurev.anthro.012809.105058

Stone, G.D. (2014) Bioinsecurity in the age of genetic engineering. In: Chen, N. and Sharp, L. (eds) *Bioinsecurity and Vulnerability*. School for Advanced Research Press, Santa Fe, New Mexico, pp. 71–86.

Stone, G.D. (2015) Biotechnology, schismogenesis, and the demise of uncertainty. *Journal of Law and Policy* 47, 29–49.

Students for Engaged and Active Learning (SEAL) (n.d.) An abbreviated history. *Divided Land, Ongoing Struggle*. Available at: https://sealstudents.wordpress.com/an-abbreviated-history/ (accessed 8 April 2016).

The Rundown Live (2013) Why march against Monsanto? (Interview wwth creator, Tami Monroe Canal). *The Rundown Live*, 19 May.

van der Ploeg, J. (2008) *The New Peasantries: Struggles for Autonomy and Sustainability in an Era of Empire and Globalization*. Earthscan, London. doi:10.4324/9781849773164

Wald, N. (2015) Towards utopias of prefigurative politics and food sovereignty. In: Stock, P.V., Carolan, M. and Rosin, C. (eds) *Food Utopias: Reimagining Citizenship, Ethics, and Community*. Routledge, New York, pp. 107–125.

Wilk, R. (2006) From wild weeds to artisanal cheese. In: Wilk, R. (ed.) *Fast Food/Slow Food: The Cultural Economy of the Global Food System*. Altamira Press, Lanham, Maryland, pp. 13–27.

3

The Use and Abuse of the Term 'GMO' in the 'Common Weal Rhetoric' Against the Application of Modern Biotechnology in Agriculture

Philipp Aerni*

Center for Corporate Responsibility and Sustainability, University of Zurich, Zurich, Switzerland

Introduction

Concerns about the risks of genetically modified organisms (GMOs) in agriculture are often framed as an ethical rather than a scientific issue. The ethical issue revolves around the question 'who are the winners and losers?' In today's debate the answer appears to be obvious: the winners are profit-seeking global companies such as Monsanto, while the losers are believed to be consumers, local farming communities and the environment that are exposed to an untested technology. Yet, after more than 20 years of experience with genetic engineering in commercial agriculture, the technology is hardly untested. In fact, new gene-editing techniques may become the next-generation breeding technologies that render the term 'GMO' obsolete.

In view of rapid technological change and industrial transformation, the public debate on GMOs, shaped by the discursive power of the opponents, seems to be increasingly anachronistic. This chapter addresses the shift of discursive power from supporters to opponents of GMOs over the past three decades from a social psychology as well as a political economy perspective. In this context, the view that opponents of GMOs are driven by ethical concerns is challenged. Opportunistic behavior by professional anti-GMO factions will be illustrated through two specific political debates on GMOs in the Swiss and the EU Parliaments. The two cases reveal how the public narrative against the case of GMOs allows opponents to conceal their private agendas behind a 'common weal rhetoric', which portrays them as selfless representatives of the common or public interest. However, as I will show, the credibility of this common weal rhetoric stands and falls with the credibility of the term 'GMO', which has developed a life of its own, very much detached from the technology itself.

The Ethical Tension Between Changing Technology and Unchanging Discursive Debate

The period between 1997 and 2017 can be characterized by the commercial expansion of GM crops in many parts of the world. In addition, new plant-breeding techniques (NBT), such as CRISPR/Cas9, have been developed that are, compared with genetic engineering, more precise, less invasive and less distinguishable from crops bred with

* E-mail: philipp.aerni@uzh.ch

established breeding techniques, including directed mutagenesis.[1] The NBT revolution is even threatening the business model of the established agroindustry based on GM crops combined with other input products. Over the same period of time, the agroecological approach to farming has become mainstream and is applied in industrial agriculture wherever it is feasible.

However, the arguments for and against GMOs today are strikingly similar to the ones used 30 years ago. Why are we not curious about recent developments? Why do we still see and hear familiar but increasingly outdated arguments for and against genetic engineering in agriculture? Wouldn't it be more effective to look at the problem on a case-by-case basis, taking into account the respective context and the technique used to address a particular challenge and possible alternatives?

While there can be many ways to answer these questions, one consideration is that shifting views is risky for political activists, especially if they have gained the moral high ground with their earlier position in the debate on agricultural biotechnology. For example, expressing general concerns about GMOs is popular with almost every constituency concerned about public health and the environment and thus allows the portrayal of oneself as an advocate for the public interest. In this common weal rhetoric, the term 'GMO' has become a convenient proxy for everything that is believed to be bad about economic and technological change in agriculture. It is relatively easy to create analogies between the presence of GMOs, loss of biodiversity, loss of traditional knowledge, loss of food sovereignty, land grabbing, farmer indebtedness, farmer suicide, environmental pollution (Aerni, 2011a) and even sterility and infant mortality (Smith, 2010). Even though these analogies may bear no direct causal relation with GMOs, they fit into a global narrative of dichotomies dividing the good and desirable from the bad and undesirable. Dichotomies help reduce complexity and provide a sense of certainty for lay people when it comes to the question of where to stand on moral issues (Heldke, 2015). As such, claims that

can be embedded in pre-existing and meaningful narratives and communicated in the form of dramatic imagery are intuitively credible to the audience, and this makes the narratives very powerful.

The powerful narrative against GMOs has the effect of making corporations, policy makers and also academic institutions anxious about becoming a target of symbolic protest once they endorse the use of GMOs in agriculture. If that happens, they are likely to become associated not only with the technology itself but also with all kinds of assumed undesirable trends associated with the term 'GMO'.

The low probability of being challenged in the public arena further strengthens the position and legitimacy of opponents of GMOs. Their influence on public opinion also translates into concrete policy outcomes that effectively ban the use of genetic engineering in agriculture, independent of the product, the market and the country context. Alas, those who suffer most from preventive regulation and the stigmatization of the technology are often not the multinational companies which have a broad portfolio of products in the business of plant protection and are able to bear the costs of burdensome and uncertain approval processes of GMO crops, but public research institutions in developed and developing countries focused on joint research on the genetic improvement of locally important orphan crops. Most were forced to abandon research that involves the use of genetic engineering due to the unwillingness of public donors to support a technology that is unpopular in their respective country.

But even as opponents of GMOs rely on simplistic, dichotomous arguments, there are internal ethical tensions within the anti-GMO coalition that arise from a growing awareness of collateral damage resulting from the stigmatization of genetic engineering in agriculture. While the anti-GMO stance may be perceived as the 'ethical' stance because it is associated with the preservation of sustainable food and agriculture, it renders almost all other technologies and inputs widely used in commercial agriculture unsustainable by default. As such, the term 'GMO' ceased to evolve as a metaphor in response to continual exchange between the abstract and the real. As a consequence, GMOs become a mere cliché in people's minds that is detached from any concrete reference to reality (Johnson, 2014). Relying on such clichés in the public debate prevents rather than enables sustainable solutions in agriculture. Such solutions are in most cases based on useful combinations of modern technologies and traditional practices that reflect the priorities of local farmers and the needs of society at large (Ogero *et al.*, 2012; Montpellier Panel, 2017). A good example may be the use of modern and affordable tissue culture laboratories by indigenous women groups in Colombia. These women have learned how to clone the locally preferred virus-free planting material and then sell the clean cassava stakes to the famers in the region for a good price. In other words, they discovered a way to add value to their traditional knowledge about the agronomic qualities of local cassava varieties by making use of modern technology.

Service-oriented agribusiness companies in the 21st century are concerned with tailor-made solutions for farmers and society that combine advanced digital and biotechnologies with agroecological approaches. For these innovative practitioners, the political struggle 'organic farming versus GMOs' is of no concern because it reflects ideologies of the 20th century that have very little to do with the actual knowledge frontier of the early 21st century.

development and commercialization of such GM crops. It owned all the major patents and aggressively lobbied abroad for permissive regulation of genetic engineering in agriculture. In this context, one could argue that the 1990s were a test ground for GM crops. The concern was real that the new technology was in the hands not only of industry but also of a specific company. Many civil society groups rightly asked for caution in embracing a new technology without appropriate public risk assessments, especially after several food contamination scandals and mad cow disease in Europe undermined trust in food safety authorities. Even though these scandals were unrelated to genetic engineering, they nevertheless increased distrust in the technology.

During this time, advocacy groups organizing anti-GMO campaigns may have contributed to a balanced public discourse on agricultural biotechnology by asking for caution and pointing to the risk of corporate power in agriculture. However, once it became clear a decade later that the media-savvy anti-GMO campaigns gained in discursive power by creating mainstream beliefs that were embraced by policy makers and powerful global retailers, the portrayal of anti-GMO activists as 'David' fighting industry as the powerful 'Goliath' no longer applied. Nevertheless, the narrative became entrenched in the mass media and social media, as well as the education system, involving visualized horror scenarios of the effects of corporate-owned GMOs in agriculture.

Transnational Advocacy Networks Against GMOs from a Political Economy Perspective

The first successfully commercialized GM varieties in the 1990s included soybean, corn, cotton and rapeseed, the major bulk commodities mainly designed for industrial purposes or animal feed. These crops consisted primarily of two simple traits: pest resistance and herbicide tolerance. The US company Monsanto took a clear lead in the

The end of deliberative demos?

DeLuca and Peeples (2002) argue that mediated information and communication technology may have replaced a deliberative demos characterized by public controversy. Political bodies, civil society groups and corporations that vie for public attention tend to focus on constructing visually appealing campaigns as part of their political and business marketing. The core objective is to deliver the message to the public in a dramatic way rather than to engage with the arguments of those who do not agree with their views. Social media may have reinforced this trend (Sunstein, 2017). In this context, the political marketing of the anti-GMO protest movement that also includes global retailers and the organic farming industry was more effective in shaping public opinion than the advertising and PR campaigns of the agro-industry (DeLuca and Peeples, 2002; Aerni and Bernauer, 2006).

From social benefits to psychological benefits

Once they reached the moral high ground in the public discourse, opponents of GMOs started to replace expensive investigative journalism with media-savvy protests based on symbols and images that equated GMOs with 'potentially catastrophic risk' and 'US interests' (Aerni and Bernauer, 2006). In other words, the public benefit of anti-GMO campaigners was no more social in nature – for example, by assuming a watchdog function – but psychological – for example, by reducing complexity and providing a simple orientation for the lay public through dichotomous portrayals of 'bad' GMO-supporting actors and 'good' GMO-opposing actors (Luhmann, 1993).

GMOs and Corporate Power

While associating the term GMO with Monsanto may have been appropriate in the 1990s, Monsanto has ceased being the unchallenged power in the business of agricultural biotechnology, because many of its GM-related patents have expired, and the company itself has been acquired by the German company Bayer. Yet, documentaries that link Monsanto to GMOs are still very popular as visual teaching material in European high schools. Moreover, the annual 'March Against Monsanto' in the USA has become a popular ritual, undeterred by the fact that Monsanto has ceased to be an exclusively American company.

Ironically, the demands to ban GM crops, indirectly meant to harm the corporate power of Monsanto, have actually harmed many of the company's smaller, more innovative competitors that are unable to comply with the increasing regulatory costs resulting from preventive regulation of GMOs. Many of these companies have developed genetically modified food products with traits that have a genuine potential to benefit society and the environment (e.g. virus-resistant cassava, drought-tolerant white maize, potato with late blight resistance, apple resistant to bacterial leaf scorch, potato with blackspot resistance and reduced borrowing due to acrylamide development). Yet, these products face continued resistance from food processing companies and global retailers that do not distinguish between incumbents and new entrants in agribusiness but merely want to ensure that their products can be labeled GMO-free. As the gatekeepers in the global food value chain that decide what will be ultimately offered to consumers in the supermarket, they actually wield more corporate power than the large and established agrochemical companies. By claiming to be 'GMO-free', they believe they are able to reassure consumers of their good motives and in turn gain their trust (Aerni, 2013).

The use and abuse of the term 'GMO'

The term 'GMO' when used in association with Monsanto is inconsistent because

- it does not cover many recombinant DNA products that belong to other areas of biotechnologies, such as 'red' (medicine) or 'white' (industrial) biotechnology;
- traits such as resistance to pests or tolerance to herbicides can often be obtained via techniques which are not technically 'GMO';
- defining what is and is not a 'GMO' is not only a scientific decision but also a political one;
- with the advance of technology, the distinction between genetic modification and other plant biotechnological techniques blurs; and

- when 'GMO' plants are processed, the results are often indistinguishable from the same 'non-GMO' products (Tagliabue, 2016).

In other words, there is no such thing as 'GMO-ness'. But the way the term is used suggests it has become a cultural construct associated with the fear of losing the natural and the innocent. In this context, the marketing of 'organic', also a cultural construct for the natural, healthy, fair and the innocent, lives from this dichotomous distinction (Blancke et al., 2015).

Is it not odd, one may ask, that butterflies killed in an organic corn field where Bt (a natural soil bacteria) is widely sprayed do not seem to produce a public outcry, but when butterflies die in a field cultivated with genetically modified Bt corn (in which the toxin of the soil bacteria is expressed in the leaves of the corn plant), there is a public outcry? The monarch butterfly has become a symbol of protest against genetic engineering in agriculture, even when studies have shown that more of them die when Bt is sprayed, since butterflies do not feed on the corn plant itself but rather on the weed (milkweed) in the corn field (Gustafsson et al., 2015; Mintz, 2017).

Similarly, Europeans do not seem to have any qualms about eating an 'organic' French baguette that is made of flour from Renan wheat, a variety popular with organic farmers that contains more transgenic material than any currently approved GMO variety (Daynard, 2015). Renan is not a product of classic plant breeding but mutagenesis induced by chemical (colchicine) or radiation – a technique that was available prior to genetic engineering and has been in use for decades (Ferrand, 2013; Johnson, 2016). So why has no one applied the metaphor 'Frankenfood' to mutagenesis? One reason might be because it would affect the 'pureness'-marketing of organic farming. Another reason might be that agriculture without crops produced by means of mutagenesis would have fatal consequences for farmers as well as consumer choice worldwide because too many of our most popular grains and vegetables are products of breeding involving techniques of directed mutagenesis. Moreover, in view of the resistance against GMOs, they have literally become the alternative to genetic engineering in industrial plant breeding (Kharkwal and Shu, 2009).

If GMO opponents are confronted with these contradictions, they tend to shift to another level of critique. Instead of discussing the technical problems associated with genetic modification, the discussion shifts to concerns about profits by large multinational agrochemical companies, the means by which the companies earn the profits (e.g. by allegedly duping farmers to adopt GM crops), and the resulting consequence of farmers becoming indebted and eventually committing suicide.

Proponents of GMOs might respond to these critiques by noting that industrial concentration is a result of costly regulation rather than ownership of GM patents (Aerni, 2014), or that research shows there is no evidence that suicide rates of farmers who adopt GM cotton in India are higher compared with those cotton farmers who have not (Gruère and Sengupta, 2011). When confronted with these facts, opponents of GM technology shift tactics again by raising the question 'who is funding you?' Even if pointing out that the studies cited are publicly funded, any response to such a question will not remove the stigma created by linking proponents to industry interests.

Efforts to maintain an artificial dichotomy between 'GMO-contaminated' and 'GMO-free' agriculture become pointless once the term 'GMO' is declared meaningless. Yet, GMO essentialism is widespread not only among the lay public but also among legal and environmental scholars who have proclaimed themselves experts on GMOs (Blancke et al., 2015). Moreover, too many actors may have a vested interest in the prevailing of GMO essentialism in the public debate (Aerni, 2011a).

Convenient dichotomies to make claims in defense of the public interest

For the French semiotician Roland Barthes, creating meaning is a product of social convention based on a shared understanding of signs that consist of the signified, an abstract concept, and the signifier that mediates the signified through images or metaphors (Barthes, 1972). As a cultural construct, the term GMO has become an empty signifier that draws its meaning from imagery language that is largely self-contained in its meaning (Clancy and Clancy, 2016). Empty signifiers serve as a way to conceal particular vested interests behind a language of universal public interest and concern (Wullweber, 2015). The very ambiguity of the term GMO demands for unambiguous, simple and identity-creating dichotomous distinctions (e.g. natural/unnatural, sustainable/not sustainable, pure/contaminated, industrial/natural) in order to reduce complexity and provide moral orientation. As such, stakeholders concerned with public legitimacy must ensure that they stand on the 'morally correct' side by adopting the common weal rhetoric against the use of GMOs in agriculture. It helps to conceal any private agenda in politics and business (Aerni, 2011b).

Internal Ethical Tensions Caused by Portraying an Ambiguous Term in an Unambiguous Way

Over the course of the past three decades, numerous vested interests in politics have emerged that benefit from the identity-creating dichotomous distinctions associated with GMOs. Sometimes it is not a material interest (as in the case of organic farming) that explains the opposition but an interest in being perceived as an actor in defense of the public interest. It represents a way to improve reputation at low cost due to the fact that an opposing statement against GMOs is hardly ever challenged in the public arena (Heldke, 2015).

GMO opponents may not have to fear changes in public perception in the near future, since the educational system has embraced a narrative of GMOs that ensures meaningfulness for the next generation (Aerni, 2013). But internal ethical tensions

may nevertheless grow due to the growing number of young, curious and critical internet users who are dissatisfied with the existing persistent stories told by like-minded groups, teachers, TV documentaries, retail marketing departments, seasoned activists and politicians. Eventually, they may notice based on their internet-based research that the claims of the health and environmental risks related to genetic engineering run counter to the findings obtained in public risk research. Hundreds of millions of US dollars have been spent in the USA and Europe on public risk assessment of GMOs. While these studies do not deny that there are real risks, they explain that these risks are not specific to genetic engineering but also occur in conventional agriculture when devoid of the basic principles of sustainable agricultural practice (EC, 2010; NAS, 2016). Recognizing that genetic engineering is already a well-tested established technique, the critical-minded may wonder why those who claim to fight GMOs in the public interest refuse to adopt a more differentiated view of modern agricultural biotechnology.

Some commentators argue that the growing social and environmental challenges facing the world in the 21st century require a combination of agroecology and advanced biotechnology (Ronald and Adamchak, 2008). To adopt a dichotomous argument – that supporters of GMOs focus merely on how to increase productivity (e.g. feed the world) while opponents care more about systemic thinking, context, culture and resilience – contradicts the fact that many new agbiotech companies, such as Simplot, Intellia, Caribou Biosciences, Cibus and Clara Foods, do not focus primarily on productivity but rather on food quality, the environment and health (The Economist, 2016). These companies combine biotech crops with big-data-driven sustainable farm management. Moreover, new breeding techniques, such as the gene-editing method called CRISPR/Cas9, have the potential to revolutionize and democratize plant breeding (Brinegar et al., 2017). This may be a reason why some leaders in organic agriculture have not declared new plant breeding techniques incompatible with organic farming (Gheysen and Custers, 2017).

from supporters to opponents of GMOs worldwide (Heller, 2001; Motta, 2014).

The shift in discursive power

The simple dichotomous narrative of 'people versus profits' does not allow for a more nuanced discussion of local challenges of agriculture and of the extent to which agricultural biotechnology may or may not be part of tailor-made local solutions. The term 'GMO' is convenient because it has the advantage of having turned into a 'meme' that can be transported in its meaning and spread in the form of images and symbols via the internet and social media (Mazanek, 2016).

After the year 2000, the anti-GMO rhetoric became part of a larger counter-hegemony narrative that comprised resistance against the World Trade Organization (WTO), US imperialism and industrial agriculture, and advocacy in favor of organic agriculture, food sovereignty and the protection of nature and indigenous rights. All of these issues underpin public concerns about genetic engineering in agriculture and why it is not good for society.

This led to a sort of scale shifting – a means by which a particular domestic discourse surrounding contentious local issues is shaped by the frame of the global counter-hegemony discourse propagated through the modern media of communication by professional international advocacy groups covering issues ranging from human rights to food sovereignty to environmentalism to eco-feminism (Della Porta and Tarrow, 2005; Tarrow, 2005). As epistemic brokers, they succeeded in establishing knowledge claims that the use of GMOs harms sustainable agriculture and offends human rights, especially indigenous rights (Herring, 2010). Yet, by taking advantage of the very limited concrete experience that consumers and taxpayers have with agricultural biotechnology, and by enhancing the diffuse anxiety about the technology through dramatic and personalized media portrayals of farmers and consumers as victims of corporate power, the discursive power has shifted

Science and the precautionary principle

An indicator of the change in public opinion and shift in discursive power from the supporters to opponents of GMOs is the regulatory response to genetic engineering in agriculture by the European Commission (EC). The EC published its communication on the precautionary principle in 2000. Taking into account not just scientific but also potential socio-economic concerns, the EC communication made it legitimate for the EU and its member states to impose bans on GMOs, even if no scientific evidence of any additional risks posed by genetic engineering have been presented. This is called the strong version of the precautionary principle. The soft version is found in the Sanitary and Phytosanitary Agreement (SPS) of the WTO. It approves a temporary ban but asks the country imposing it eventually to produce scientific evidence of risk, otherwise the ban has to be lifted again.

In consideration of the soft version of the precautionary principle, the WTO dispute panel on GMOs decided in 2005 to support the plaintiffs (United States, Canada and Argentina) by arguing that the approval process for GMOs in the EU causes undue delay. The decision was not appealed by the EU. The plaintiffs would therefore have been authorized to undertake retaliation measures (e.g. punitive tariffs on selected European imports) (Bernauer and Aerni, 2009). Yet, they refrained from doing so in the hope that the EU will eventually align its regulatory framework on GMOs with WTO requirements. So far the EC responded only with a regulatory reform that contains an opt-out clause for member states, and a promise to review the existing regime of regulating not only GMOs but also new breeding techniques such as gene editing (EC, 2015). As a result, many member states have legalized their bans of GM crops that have been approved at the EU level for cultivation.

The lack of political will of the EC to challenge the prevailing popular narrative on the potentially catastrophic risks of GMOs manifested itself once again when Jean-Claude Juncker, the President of the EC, decided in 2015 to abolish the post of Science Advisor to the EC in response to GMO opponents asking for the dismissal of Anne Glover, who failed to acknowledge the dangers of GMOs (Wildson, 2014). These decisions, combined with ever more restrictive regulation on field trials and the commercial release of GM crops, strengthen the view of the European public that there must be something wrong with the technology. In other words, anti-GMO advocacy groups have succeeded not only in shaping public opinion in Europe but also political decision-making processes on genetic engineering in agriculture. This success in gaining public legitimacy and political power may have led to political opportunism that will be illustrated in two case studies of events in the parliaments of Switzerland and the European Union.

Case 1: Ethical Concern or Political Opportunism in Switzerland?

In 2006, the Swiss Federal Council launched a national research program on the risk and benefits of genetically modified crops, called NRP59. One of the main causes for funding this type of research through the Swiss National Science Foundation was the approval in 2005 of a national referendum for a five-year ban on the cultivation of genetically modified crops.

In view of the adverse attitude toward the technology, it was surprising that politicians of the main political parties (including the Green Party) emphasized the importance of science in political decision making after approving the referendum. As such, regulation of GMOs after the end of the moratorium would be contingent on the findings of NRP59. In 2013, after a first three-year extension of the moratorium, the results of numerous NRP59 projects were published at www.nfp59.ch. The summary report admitted that there are risks related to the cultivation of GM crops in Switzerland, but that these

risks would also be found in conventional agriculture. They are related to inadequate sustainable practices and not necessarily to the GM technology itself. The Swiss parliament was aware of these findings before they were officially presented in September 2013. In order to avoid an inconvenient debate, the MPs decided in August 2012 to extend the moratorium for another three years. The Swiss National Science Foundation expressed its disapproval regarding the deliberate avoidance of debate in view of the prior agreement of the politicians to take the results of the NRP59 seriously.

In 2016, the federal council found that public acceptance would be insufficient to lift the ban on GM cultivation in Switzerland. It submitted a report to the parliament citing a study on revealed consumer behavior of GM foods in Switzerland, which was part of NRP59, that allegedly (though, we will see, incorrectly) provided evidence that Swiss consumers are unlikely to buy genetically modified foods (Aerni et al., 2011). The study tested revealed consumer behavior by providing Swiss consumers with the freedom of choice between three types of clearly labeled corn bread, one made with organic corn, one with conventional corn and one with genetically modified corn, respectively. Moreover, the study compared revealed political preferences (voting decision in the referendum in 2005) and revealed consumer preference (purchase at the market stand).

The discrete choice model used for the field study with market stands allowed researchers to measure sensitivities related to different price scenarios, sales groups, locations and package sizes. The corn bread was sold at five locations in four major cities in the French and German part of Switzerland. Roughly 5000 loaves of bread were sold to 3000 consumers and 1000 questionnaires were returned (found in the bread bag for the customer). It turned out that sales increased at all market stands on average by 30% once the GM option was available. Moreover, the package size (small/big bread) was more important than the product type (GM, conventional, organic) in explaining consumer behavior, and the price hardly mattered. The share of consumers who decided to buy GM corn bread was above 20% even if it was as expensive as organic. In other words, behavior toward GM corn bread was not any different from behavior toward any other novel food product introduced in the market. Moreover, voting behavior and purchasing behavior were not consistent with consumer decisions to purchase bread.

In the parliamentary debate on the first extension of the moratorium, the results of the study were simply ignored. But in the documentary material in favor of the second moratorium prepared by the federal council for the parliament, the results were portrayed in a deliberately misleading way, suggesting that the results of the field study would provide evidence that Swiss consumers would not buy genetically modified food. When the author of the study wrote an op-ed in the national daily *Tages-Anzeiger*, titled 'Politics as the enemy of science', denouncing the unsupported portrayal of the study, the president of the parliamentary commission on science, education and culture (WBK) replied by simply denying it, without citing any study that would confirm her view that consumers would avoid GM foods. Oddly, it was this commission that proposed to the parliament to reject the request of the federal council to extend the temporary moratorium and instead asked for a permanent moratorium. Why? The commission argued that there is a lack of willingness among its members to review the state of the art in risk and benefit research on GM crops every four years again. The open admission of the commission that they are not really interested in science and the misleading portrayal of the results of empirical research funded through NFP59 raises the question to what extent the opponents have actually abandoned their commitment to proper ethical conduct not just in Switzerland but in the EU as well.

Case 2: The Heubuch Report of the EU Parliamentary Commission on Development

As in the case of Switzerland, politicians representing national socialist and green parties in Europe were eager to chair commissions in the European Parliament that deal with the issue of GMOs. The Rapporteur of the Committee on Development of the European Parliament, Maria Heubuch, is from the German Party Bündnis/die Grünen. Her career as a national politician with a farming background in Germany can be characterized by her lobbying activities for agricultural trade protection and against industrial agriculture. As a classical farm lobbyist in the German state of Baden Würtemberg, she framed her private interest in the continuation of farm subsidies as a public interest issue, namely to protect sustainable and traditional agriculture in Germany against the power of global agri-business. By joining the Green Party, she ensured that her political activities were not called lobbying but actually 'advocacy work', which also included opposition against GMOs.

In March 2016, Heubuch presented the Report of the Committee on Development on the New Alliance for Food Security and Nutrition (NAFSN) to the European Parliament. NAFSN was an initiative launched in 2012 by the G8 to mobilize private-sector investment for African agriculture in collaboration with African governments. The content of the Heubuch report was widely celebrated by anti-corporate, anti-GMO and anti-G8 activists.

After first citing international reports and agreements on sustainable development, food security and the environment that were endorsed by the European Union, the report portrays the NAFSN as an initiative that runs counter to prior efforts to promote sustainable agriculture, environmental protection and food security in Africa. After all, NAFSN involves the private sector and, even worse, does not necessarily exclude the use of agricultural biotechnology.

The report stands out for its paternalistic tone, portraying African governments participating in the initiative through the adoption of country cooperation frameworks as ignorant, gullible or worse, deliberately selling their country's land to multinational corporations at the expense of their own small-scale farmers. Moreover, the report warns European donors involved in the initiative of becoming complicit in a plot to replicate a second green revolution in Africa at the expense of sustainable agriculture. It reminds

EU member states of their commitments to multilateral environmental agreements such as the Convention on Biological Diversity that would essentially ban the use of GMOs through a strong version of the precautionary principle entrenched in its Cartagena Protocol on Biosafety.

The report does not mention the intense pressure of European public and private donors and NGOs on African governments to stay away from GMOs over the past two decades (Paarlberg, 2009). African countries knew they would be on the safe side in securing European aid and market access for their agricultural goods if they adopt a template of the European biosafety regulation, which was effectively declared as dysfunctional by the WTO when it comes to the approval process of GMOs.

Apart from urging G8 member states that support NAFSN not to support GMOs in Africa, the report criticizes the spread of certified seeds in Africa that happen through NAFSN. What is wrong with certified seeds? One study highlighted that the average adoption rate of improved hybrid maize seed was 44% and increasing fast in East Africa (Marechera et al., 2016). This high adoption rate has neither prevented farmers who want to use farm-saved seed from doing so nor undermined seed diversity. The challenge with traditional small-scale farming in Africa is that, due to high fertility rates and lack of off-farm employment, average farm sizes have been shrinking over the past decades in many rural areas of Kenya and Ethiopia (Aerni 2015a). Shrinking farms have been a cause of malnutrition and starvation, or otherwise deforestation and migration in rural Europe in the 19th century, and they are so today in rural Africa.

In this context, a blog posting by Margret Karembu (2016), who grew up on a farm in Africa, accuses the Heubuch report for its unrealistic pretensions. According to her, rather than telling farmers that they should stick to traditional farming while accepting agroecology consultants from Europe, Europeans should give farmers the freedom of choice. She finds it cynical that a European left-wing politician advocates exclusive support for small-scale farming in Africa while simultaneously endorsing costly private and public food standards for food imports to Europe that make it virtually impossible for marginal small-scale farmers in Africa to benefit from European market access. She also points out that Europe has approved at least 86 GM crop products and imported more than 30 million tons of GM soy bean for use as animal feed, making the claim to protect Africans from the risks of GM crops look rather hypocritical.

By comparing NAFSN with the Green Revolution, the Heubuch report reveals a lack of understanding of history. The Green Revolution was a public-sector initiative, not a public–private partnership (Aerni, 2015b). The report also refers to prior accords, reports and initiatives that are assumed to be in line with its claims. But several are not: many parts of the Convention on Biological Diversity, the Comprehensive Africa Agriculture Development Programme (CAADP) and the Busan Declaration on Partnership for Effective Development Cooperation (BDPED) run counter to her claims.

The Convention on Biological Diversity states in Article 16 that 'each Contracting Party, recognizing that technology includes biotechnology, and that both access to and transfer of technology among Contracting Parties are essential elements for the attainment of the objectives of this Convention'. The Convention further urges Parties in Article 19 to promote priority access to the benefits arising from biotechnologies, especially for developing countries. A blog response by Diran Makinde (2016) denounced the report for being in breach of the Convention on Biological Diversity. He also wonders why the report limits its demand to stick to small-scale subsistence farming to Africa only. Why isn't it extended to other developing countries? Does Europe still consider Africa its back yard? This would reveal the neocolonial underpinnings of the report.

The CAADP was endorsed by the African Union and the New Partnership for African Development (NEPAD) in 2003. CAADP is an Africa-led and Africa-owned initiative and framework to rationalize and revitalize African agriculture for economic growth and lasting poverty reduction. It champions reform in the agricultural sector by setting two main targets among its members that are reviewed on a regular basis: 6% annual growth in agricultural GDP and an allocation of at least 10% of public expenditures to the agricultural sector. CAADP emphasized the importance of private-sector investments (NEPAD, 2011). Yet, the Heubuch report implies that this Africa-owned initiative wants merely to preserve small-scale subsistence agriculture.

According to the BDPED, one of the main 'shared' principles of the BDPED and the prior OECD Paris Declaration of Aid Effectiveness is 'ownership of development priorities by developing countries: Countries should define the development model that they want to implement' (OECD, 2011, n.p.). CAADP expressed the Africa-owned development priority of agricultural modernization, but the Heubuch report seems to claim otherwise. The Heubuch report, which openly urges donors to refrain from supporting agricultural modernization in Africa, nevertheless refers to the BDPED ('having regard to...'), as if the content of the report is in line with the 'shared principles'.

One could argue that Maria Heubuch and her committee were simply ignorant of the content of these accords and declarations. Yet, if they were aware of the content, then one has to question the ethical conduct of the parliamentary committee on development. The European Parliament itself is hardly able and willing to scrutinize the quality and ethics behind such committee reports. Moreover, civil society groups that usually assume the role of independent watchdogs may be too concerned about upsetting their own constituency by highlighting ethical misconduct by members of the European Parliament that use their own common weal rhetoric against GMOs.

Concluding Remarks

The term 'GMO' has been embraced in law and politics in an essentialist way that allows proponents and opponents to make simple and dichotomous distinctions that reduce complexity and, simultaneously, provide meaning, identity and orientation. In public narratives that rely on GMO essentialism, GM technology is associated with negative attributes, such as unnatural, unsustainable, contaminated, risky and unfair, whereas alternatives such as organic farming and agroecological approaches are linked with the corresponding positive attributes. The narrative is embedded in the popular ahistorical myth of the original farming community that was in harmony with nature and societal needs. This state of blissful equilibrium is then disrupted by technological and economic change, an external force imposed on the community. The agents of this force are profit-seeking multinational companies. They try to coax innocent local farmers into adopting genetically modified crops to generate profits at the expense of the local environment and community. Fortunately, anti-GMO activists and their allies in politics, the mass media, the retail industry and the organic farming industry come to the rescue by seeking to ban the use of GMOs and restore the old harmony with nature and the community and by introducing sustainable agriculture – understood as organic farming or agroecological improvements of local agricultural systems. The sustainable agricultural products cultivated by the restored farming community are then sold to affluent consumers who are invited to join the epical struggle against bad industrial agriculture by paying a small price premium. This mythical story is timeless and strongly attached to the values of postmaterial societies (Rangan, 2001). It has therefore been integrated into a common weal rhetoric by many stakeholders in affluent economies who would like to be seen as actors on the 'good' side of the epical struggle. By making use of this common weal rhetoric they strengthen their public legitimacy, and may indirectly gain votes and/or new customers at low cost.

However, the success of this convenient way of gaining popularity while concealing the concrete private agenda behind the claim to act on behalf of the public interest stands and falls with the term 'GMO'. The term is used as a metaphor in the mythical struggle against industrial agriculture but is increasingly detached from the concrete reality of the adoption of certain GM crops and its consequences, as well as from the fact that agricultural biotechnology covers a wide range of techniques, products and processes that offer context-specific challenges and opportunities when applied in agriculture. As such, the use and abuse of the term 'GMO' may also produce considerable collateral damage for society and the environment over the long run because it excludes in advance any potential positive impact of the technology for sustainable agriculture.

The global sustainability challenges in agriculture cannot be effectively addressed by either relying on GMOs or on organic alone. Effective solutions must be based on locally tailored combinations of advanced technologies as well as system-oriented organic or agroecological farming practices. It should, however, be the local farmers rather than the input industry, retail industry or anti-GMO advocates who decide which combination is most sustainable for their particular local context. Alas, the narrative of an epical global struggle for or against GMOs deprives local stakeholders of their own local stories and the articulation of their own local interests (Barthes, 1972; Rangan, 2001).

The two cases presented here illustrate the collateral damage of the common weal rhetoric relying on the use and abuse of the term GMO. Despite the inaccurate claims and the paternalistic tone, the anti-GMO narrative prevails in Europe and has entrenched itself also in the education system. Yet, the dichotomous portrayals in the GMO debate make sense to the public only as long as the term GMO makes sense as a metaphor for everything unwanted in agriculture. New gene-editing techniques such as CRISPR/Cas9, as well as successful combinations of the agroecological approach with agricultural biotechnology, may eventually challenge the old metaphor and open space for effective collaboration and tailor-made solutions for farmers. This will also be the moment when the established common weal rhetoric ceases to be perceived as the 'ethical' view and the role of science and empirical research may again gain in importance in policy decision making.

Note

[1] The variation created by natural mutations has been the basis for all plant breeding since humans started cultivating plants thousands of years ago. Starting in the 1930s, the use of radiation and chemicals to induce mutations became possible. This is a random approach, because mutations have to be induced on thousands of seeds and then only those rare beneficial mutations are selected. For Oligonucleotide-Directed Mutagenesis, a complementary nucleotide sequence is used to introduce a mutation at a very specific location in the genome (see Epso, 2016).

References

Aerni, P. (2011a) Food sovereignty and its discontents. *African Technology Development Forum (ATDF) Journal* 8, 23–40. Available at: http://atdforum.org/atdf-journal-food-sovereignty/ (accessed 1 November 2017).

Aerni, P. (2011b) Die Moralisierung der Politik als Kehrseite der Angst vor dem globalen Wandel. In: Aerni, P. and Grün, K.-J. (eds) *Moral und Angst*. Vandenhoeck und Ruprecht Verlag, Göttingen, Germany, pp. 13–32.

Aerni, P. (2013) Resistance to agricultural biotechnology: the importance of distinguishing between weak and strong public attitudes. *Biotechnology Journal* 8, 1129–1132. doi:10.1002/biot.201300188

Aerni, P. (2014) The motivation and impact of organized public resistance against agricultural biotechnology. In: Castle, D. *et al.* (eds) *Handbook on Agriculture, Biotechnology and Development*. Edward Elgar, Cheltenham, UK, pp. 482–521.

Aerni, P. (2015a) *The Sustainable Provision of Environmental Services: from Regulation to Innovation*. Springer Series on CSR, Ethics and Governance. Springer, Dordrecht, Netherlands.

Aerni, P. (2015b) Agricultural biotechnology and public attitudes: An attempt to explain the mismatch between experience and perception. In: Watson, R. and Preedy, V.R. (eds) *Genetically Modified Organisms in Food*. Elsevier, Amsterdam, pp 149–156.

Aerni, P. and Bernauer, T. (2006) Stakeholder attitudes towards GMOs in the Philippines, Mexico and South Africa: The issue of public trust. *World Development* 34, 557–575. doi:10.1016/j.worlddev.2005.08.007

Aerni, P., Scholderer, J. and Ermen, D. (2011) What would Swiss consumers decide if they had freedom of choice? Evidence from a field study with GM corn bread. *Food Policy* 36, 830–838. doi:10.1016/j.foodpol.2011.08.002

Barthes, R. (1972) *Mythologies*. Translation by Annette Lavers. Noonday Press, New York.

Bernauer, T. and Aerni, P. (2009) Trade conflict over genetically modified organisms. In: Gallagher, K. (ed.) *Handbook on Trade and Environment*. Edward Elgar, London, pp. 184–194.

Blancke, S., Van Breusegem, F., De Jaeger, G., Braeckman, J. and Van Montagu, M. (2015) Fatal attraction: The intuitive appeal of GMO opposition. *Trends in Plant Science* 20, 414–418. doi:10.1016/j.tplants.2015.03.011

Brinegar, K., Yetisen, A., Choi, S., Vallillo, E., Ruiz-Esparza, G.U., Prabhakar, A.M. and Yun, S.H. (2017) The commercialization of genome-editing technologies. *Critical Reviews in Biotechnology* 37, 1–12. doi:10.1080/07388551.2016.1271768

Clancy, K.A. and Clancy, B. (2016) Growing monstrous organisms: The construction of anti-GMO visual rhetoric through digital media. *Critical Studies in Media Communication* 33, 279–292. doi:10.1080/15295036.2016.1193670

Daynard, T. (2015) A genetically modified organic wheat? It already exists. Terry Daynard's Blog. Available at: https://tdaynard.com/2015/12/12/a-genetically-modified-organic-wheat-it-already-exists/ (accessed 1 November 2017).

DeLuca, M.K. and Peeples, J. (2002) From public sphere to public screen: Democracy, activism, and the 'violence' of Seattle. *Critical Studies in Media Communication* 19, 125–151. doi:10.1080/07393180216559

Della Porta, D. and Tarrow, S.G. (eds) (2005) *Transnational Protest and Global Activism*. Rowman & Littlefield, Lanham, MD.

Epso (2016) Oligonucleotide-directed mutagenesis: Matchmaking and single mismatching (crop genetics improvement techniques fact sheet). Available at: http://www.epsoweb.org/file/2182 (accessed 24 March 2018).

European Commission (EC) (2010) A decade of EU-funded GMO research. EUR 24473 EN: European Commission, Brussels. Available at: http://ec.europa.eu/research/biosociety/pdf/a_decade_of_eu-funded_gmo_research.pdf (accessed 1 November 2017).

European Commission (EC) (2015) More freedom for Member States to decide on the GMOs use for food and feed. Press release, 22 April. Brussels, Belgium. Available at: http://europa.eu/rapid/press-release_IP-15-4777_en.htm (accessed 1 November 2017).

Ferrand, E. (2013) Le blé préféré de l'agriculture biologique. Available at: http://emmanuelferrand.blogspot.ch/2013/02/le-ble-prefere-de-lagriculture.html (accessed 1 November 2017).

Gheysen, G. and Custers, R. (2017) Why organic farming should embrace co-existence with cisgenic late blight-resistant potato. *Sustainability* 9, 172. doi:10.3390/su9020172

Gruère, G. and Sengupta, D. (2011) Bt cotton and farmer suicides in India: An evidence-based assessment. *Journal of Development Studies* 47, 316–337.

Gustafsson, K.M., Agrawal, A.A., Lewenstein, B.V. and Wolf, S.A. (2015) The monarch butterfly through time and space: The social construction of an icon. *BioScience* 65, 612–622. doi:10.1093/biosci/biv045

Heldke, L. (2015) Pragmatist philosophical reflections on GMOs. *Journal of Agricultural and Environmental Ethics* 28, 817–836. doi:10.1007/s10806-015-9569-4

Heller, C. (2001) From risk to globalization: Discursive shifts in the French debate about GMOs. *Medical Anthropology Quarterly* 15, 25–28. doi:10.1525/maq.2001.15.1.25

Herring, R.J. (2010) Epistemic brokerage in the bio-property narrative: contributions to explaining opposition to transgenic technologies in agriculture. *New Biotechnology* 27, 614–622. doi:10.1016/j.nbt.2010.05.017

Johnson, N. (2014) What I learned from six months of GMO research: None of it matters. *Grist*, 9 January. Available at: http://grist.org/food/what-i-learned-from-six-months-of-gmo-research-none-of-it-matters/ (accessed 1 November 2017).

Johnson, N. (2016) Es ist fast unmöglich zu definieren, was Gentechnik ist. *Krautreporter*, 21 January. Available at: https://krautreporter.de/1265--es-ist-fast-unmoglich-zu-definieren-was-gentechnik-ist (accessed 1 November 2017).

Karembu, M. (2016) How European-based NGOs block crop biotechnology adoption in Africa. Available at: https://geneticliteracyproject.org/2017/02/23/european-based-ngos-block-crop-biotechnology-adoption-africa/ (accessed 1 November 2017).

Kharkwal, M.C. and Shu, Q.Y. (2009) The role of induced mutations in world food security. In: Shu, Q.Y. (ed.) *Induced Plant Mutations in the Genomics Era*. Proceedings of a 2008 International Joint FAO/IAEA Symposium, Vienna, Austria, pp. 33–38.

Luhmann, N. (1993) *Risk: A Sociological Theory*. Walter de Gruyter, Berlin.

Makinde, D. (2016) Institutionalizing poverty in Africa by members of the European Parliament.

Available at: http://atdforum.org/institutionalizing-poverty-in-africa-by-mep/ (accessed 1 November 2017).

Marechera, G., Muinga, G. and Irungu, P. (2016) Assessment of seed maize systems and potential demand for climate-smart hybrid maize seed in Africa. *Journal of Agricultural Science* 8, 171–181. doi:10.5539/jas.v8n8p171

Mazanek, C. (2016) Frankenfoods: Conceptualizing the anti-GMO argument in the Anthropocene. *New Errands: The Undergraduate Journal of American Studies* 3, 1–12.

Mintz, K. (2017) Arguments and actors in recent debates over US genetically modified organisms (GMOs). *Journal of Environmental Studies and Sciences* 7, 1–9. doi:10.1007/s13412-016-0371-z

Montpellier Panel (2017) Agriculture for impact: Tissue culture. Available at: http://ag4impact.org/sid/genetic-intensification/biotechnology/tissue-culture/ (accessed 1 November 2017).

Motta, R. (2014) Social disputes over GMOs: An overview. *Sociology Compass* 8, 1360–1376. doi:10.1111/soc4.12229

National Academies of Sciences, Engineering, and Medicine (NAS) (2016) *Genetically Engineered Crops: Experiences and Prospects.* NAS, Washington, DC. doi:10.17226/23395

NEPAD (2011) CAADP engagement to call for improved private sector partnerships and investments. Available at: http://www.nepad.org/content/caadp-engagement-call-improved-private-sector-partnerships-and-investments (accessed 1 November 2017).

OECD (2011) The Busan Partnership for Effective Development Co-operation. Available at: http://www.oecd.org/development/effectiveness/busanpartnership.htm (accessed 1 November 2017).

Ogero, K.O. *et al.* (2012) *In vitro* micropropagation of cassava through low cost tissue culture. *Asian Journal of Agricultural Sciences* 4, 205–209.

Paarlberg, R. (2009) *Starved for Science.* Harvard University Press, Cambridge, Massachusetts. doi:10.4159/9780674041745

Rangan, H. (2001) *Of Myth and Movements: Rewriting Chipko into Himalayan History.* Verso Books, New York.

Ronald, P.C. and Adamchak, R.W. (2008) *Tomorrow's Table: Organic Farming, Genetics, and the Future of Food.* Oxford University Press, Oxford.

Smith, J. (2010) Genetically modified soy linked to sterility, infant mortality in hamsters. *The Huffington Post.* Available at: https://www.huffingtonpost.com/jeffrey-smith/genetically-modified-soy_b_544575.html (accessed 1 November 2017).

Sunstein, C. (2017) *#Republic: Divided Democracy in the Age of Social Media.* Princeton University Press, Princeton, New Jersey.

Tagliabue, G. (2016) The necessary 'GMO' denialism and scientific consensus. *Journal of Science Communication* 15, 1–11. Available at: https://jcom.sissa.it/sites/default/files/documents/JCOM_1504_2016_Y01.pdf (accessed 29 June 2018).

Tarrow, S. (2005) *The New Transnational Activism.* Cambridge University Press, Cambridge, UK.

The Economist (2016) The future of agriculture. *Technology Quarterly,* 9 June. Available at: http://www.economist.com/technology-quarterly/2016-06-09/factory-fresh (accessed 1 November 2017).

Wildson, J. (2014) Juncker axes Europe's chief scientific adviser. *The Guardian,* 13 November. Available at: http://www.theguardian.com/science/political-science/2014/nov/13/juncker-axes-europes-chief-scientific-adviser (accessed 1 November 2017).

Wullweber, J. (2015) Global politics and empty signifiers: The political construction of high technology. *Critical Policy Studies* 9, 78–96. doi:10.1080/19460171.2014.918899

4

Collaborating with the Enemy? A View from Down Under on GM Research Partnerships

Rachel A. Ankeny,* Heather J. Bray and Kelly A. McKinley

School of Humanities, University of Adelaide, Australia

Introduction

The introduction of genetically modified (GM) crops and food has generated long-running and often polarized public debate, and the fact that ethical tensions exist about GM agriculture is undeniable. Although many commentators have reflected on public concerns associated with 'changing nature', possible risks from GM technologies and the involvement of large multinational corporations (Thompson, 2007; Ankeny and Bray, 2018), there has been less attention in the scholarly literature on the role of public–private partnerships in GM research.

This chapter focuses on public–private funding patterns and partnerships in the development of GM crops and foods in the Australian context over the past two decades. GM research and development (R&D) processes have several ethical tensions associated with them: one key issue is who gains or profits from GM research and products. Many people who are not opposed to GM research in principle fear that when private entities are involved, particularly the large multinationals with which GM research is frequently associated, shortcuts will be taken in the name of profits, resulting in increased risks to human health and/or the environment associated with the work. Others question why such research is worth pursuing when the benefits are primarily associated with commercial needs and goals, rather than public benefits such as addressing global food security. More generally, there are concerns that the values traditionally associated with public research (such as openness, data sharing, transparency and public benefit) are in fundamental conflict with those underlying privately funded research (where commercial benefit and protection of intellectual property are critical).

Australia is an ideal locale for this exploration, given that industry-funded research has increased significantly in recent years. Australia also provides a useful case study as it does not have complete bans on GM crop growth or use in the food supply (unlike parts of the EU, for example, until recently) and GM products are not widespread (unlike the USA). In addition, there have not been detailed studies about public–private collaborations in the Australian context. Furthermore, public opinions on and regulatory approaches to GM in Australia remain mixed, representing various tensions that exist in attitudes toward GM.

In order to shed light on the ethical tensions noted above, we use a quantitative data analysis of applications to the Australian regulatory authority for intentional release of a genetically modified organism (GMO) to explore the actual distribution of public, private and other forms of funding underlying the research, and patterns associated with types

* Corresponding author. E-mail: rachel.ankeny@adelaide.edu.au

of crops and traits modified, in order to show that the typical patterns that have previously been found elsewhere are not the case in the Australian context. In addition, we develop short case studies based on publicly available information, grey and published literature, and regulatory data in order to promote deeper reflection on the supposed public–private divide in research and to emphasize the need for scholars to explore the complex partnerships that often underlie GM research. Although a highly detailed analysis is not possible given the available data, we contend that the Australian setting provides a different perspective on the potential for various forms of public–private collaborations in GM research, as well as an excellent test bed for assessing effects of diverse types of funding and institutional arrangements. We use this analysis to illuminate some critical issues related to better understanding the tensions associated with this type of research and elucidate issues that require additional attention.

Background: GM and Public–Private Collaborations in Australia

Early development of GM food plants in Australia occurred in step with other regions, including Europe and the USA, although Australia created one of the earliest oversight bodies focused on GM based on voluntary guidelines (Australian Academy of Science, 1980). In 1987, the first GMO was released outside the laboratory (a GM agrobacterium, later commercialized as 'No Gall') and was the third recombinant DNA organism in the world to be field tested (Hindmarsh, 2008; Kerr, 2011). The first commercial release of a GM plant in Australia, a blue carnation for floriculture (Lu et al., 1991), occurred in 1995 (GMAC, 1996). The first GM crop, released in 1996 (GMAC, 1997), was an insect-resistant cotton (Cousins et al., 1991) known as Bt or Ingard cotton, and was developed by the Australian public agency Commonwealth Scientific and Industrial Research Organisation (CSIRO) using a gene owned by Monsanto and licensed by CSIRO, in partnership with Cotton Seed Distributors (Davidson, 2003).

The Commonwealth *Gene Technology Act 2000* came into effect on 21 June 2001 (Hain et al., 2002) due to increasing concern among the public and lack of adequate information to help people make informed decisions, perceptions that industry could not be relied on to be sufficiently rigorous, and the need for transparency and a uniform regulatory system. Its aim is 'to protect the health and safety of the people, and to protect the environment, by identifying risks posed by or as a result of gene technology and by managing those risks through regulating certain dealings with GMOs' (Commonwealth of Australia, 2001, section 3). Since the Act, the responsibility for regulating GMO dealings has rested with the Office of the Gene Technology Regulator (OGTR), which primarily relies on peer-review processes performed by professional scientists. The Act explicitly excludes economic and social arguments, despite calls from critics to revise the legislation to require their consideration during licensing (Wickson, 2007; Hindmarsh, 2008). Other scholars have criticized the required scientific assessment as too narrow in scope, arguing that the definition of the 'environment' does not include ecosystem analysis (Lawson, 2002), and that the system facilitates approval of GM foods that is too rapid (Levidow and Carr, 2000) and thus is regulation *for* industry than regulation *of* industry (Lockie et al., 2005).

Australia is currently ranked twelfth in the world in terms of the area of land sown with GM crops (ISAAA, 2016), particularly cotton and canola. However, relatively few GM crops have been approved for commercial release in Australia, compared with other countries (such as the USA); the total number of applications made to the OGTR for licenses on an annual basis is small. Nonetheless, GM crops remain a highly controversial issue in Australia. Shortly after OGTR approval of the commercial release of InVigor canola in 2003, bans on growing GM food crops were established in canola-growing states. These moratoria have been attributed to various anti-GM campaigns and state-based political issues, including concerns about the potential economic impacts of GM canola on Australia's access to export markets where GMOs are not permitted or are greatly limited, such as Japan and the European Union (Hindmarsh, 2008; Tribe, 2012). By 2010, GM canola was permitted (with some restrictions) in most states. Currently, general bans only persist in Tasmania and South Australia (where the moratorium was recently extended to 2025).

Australians are generally considered to be less cautious about GM than Europeans and more hesitant than those in the USA. Yet, most studies of Australian consumers have found that attitudes to biotechnology in food production, including GM foods, tend to be more negative than positive (Bray and Ankeny, 2017) and the adequacy of GM food labeling in Australia is contested (Bray and Ankeny, 2015). Moreover, previous qualitative research has shown that the purpose for which GM is used and who will benefit from it are critical to Australian consumers' views on GMOs (Ankeny and Bray, 2016). Direct anti-GM activism has been far more limited in Australia compared with in Europe or the USA; the 2011 destruction of a CSIRO field trial of GM wheat with altered nutritional value represents an extreme form of protest for Australia (described below in case study 2). Popular concern continues about the use of GM in crops destined for the food supply and the potential for drift between GM and non-GM crops (especially organics), highlighted in a recent court case in Western Australia (Neales, 2013).

GM research originally has occurred within an innovation system in Australia that was historically characterized as having a 'low level of science and technology expenditure, a high level of government involvement in financing and undertaking research, a low level of private sector research and development and exceptionally high dependence on foreign technology' (Gregory, 1993, p. 324). Efforts were made in the 1990s and early 2000s to increase industry contributions to academic research, via tax and other incentives for collaborative research (Collier, 2007). For instance, the Australian Cooperative Research Centres (CRCs), which began in 1990, are multisite collaborative R&D ventures bringing together university and public-sector research and promoting the flow of

knowledge and technical skills between private industry and public organizations (Encyclopedia of Australian Science, 2011). The 2001 Commonwealth package *Backing Australia's Ability* provided significant support for the commercialization of research conducted in universities and publicly funded research agencies. The National Competitive Grants Program of the Australian Research Council (ARC) instituted Linkage schemes in the early 2000s that are intended to encourage collaborative research especially with industry. Despite these efforts, Australia is still claimed to underperform based on most measures of collaboration (Commonwealth of Australia, 2010).

Conflicts Created by Public–Private Collaborations?

Traditionally, academic research and industrial/commercial research have long been interdependent. However, recent commentators have decried the negative influences of industry on academic science, which they see as having increased in recent times. Indeed, contemporary critiques contend that partnerships between industry and universities have become more varied, aggressive and publicly visible, and wider in scope in recent years (e.g. Lacy *et al.*, 2014).

Although considerable resources are necessary to pursue many forms of modern research, money from industry often is viewed with suspicion, even by scientists themselves (Biscotti *et al.*, 2009). Criticisms range from potential to compromise the research problem choice and priority setting, to falsification or suppression of research results to suit commercial interests, particularly in the biomedical sector (e.g. Krimsky, 2003; Sismondo, 2008; Elliott, 2010). Some contend that the internal norms of academic science and commercial research are in principle incompatible (e.g. Blumenthal *et al.*, 1996; Slaughter and Leslie, 1997; Krimsky *et al.*, 1999; Bok, 2003), as profit motivations necessarily run counter to traditional academic values associated with scientific inquiry and free flow of knowledge and information (e.g. Hackett, 2005). Others contend that these arguments require a blind adherence to existing systems without adequately considering that industry–academia–governmental, or so-called 'triple helix' (Etzkowitz and Webster, 1998), collaborations may represent new modes of knowledge production (Gibbons et al., 1994). More recently, scholars have noted that the boundaries between academic and commercial research are no longer fixed or rigid, as these domains are increasingly interwoven, especially in medical and agricultural biotechnologies (Vallas and Kleinman, 2008; Kleinman, 2010). Thoughtful scholarship has emerged about this interplay as well as potential regulatory and other forms of solutions (e.g. Radder, 2010).

Agricultural biotechnology is relevant for general debates on public–private collaborations in science, because agriculture was one of the earliest fields that attracted significant commercial investment (Busch et al., 1991). Agricultural biotechnology has also traditionally received considerable public investment in the USA, Australia and elsewhere, with substantial efforts to attract industry funds for research collaborations (Mowery et al., 2004). Our springboard for this chapter's analysis is Welsh and Glenna's (2006) study showing that US university research on transgenic crops has increasingly mirrored the research profile of for-profit firms during the period 1993–2002 and that private-sector firms have dominated R&D and commercialization processes for GM. Welsh and Glenna conclude that these trends have led to a narrowing focus on a few commercially important crops with plant-protection traits such as herbicide tolerance (HT) and insect resistance (IR). The reason is because these allow for substantial returns on R&D investments (Ervin et al., 2001) and are linked to agricultural inputs produced and sold by the same companies (e.g. glyphosate-based herbicides). In contrast, critics note that staple food crops (FAO, 2004) and other traits, such as those with environmental or nutritional benefits, have been relatively neglected, especially during the early years of GM research. Some argue that to correct this situation, the public sector must pursue a greater share of transgenic crop development via direct funding or

financial incentives, with emphasis on traits associated with publicly valued benefits (e.g. Doering, 2004), or even less commercially relevant or subsistence crops, sometimes termed 'orphan' crops (e.g. Paarlberg, 2000).

In addition, Lacy and collaborators' recent survey (2014) of US university- and industry-based scientists and others participating in agricultural biotechnological research collaborations identified concerns over their 'distinct cultures' with different values and goals, understandings of their research environments and criteria for research agenda choice. Although their one shared criterion for problem choice was the public good, it is noted that underlying this concept might be fundamentally different ideas, given that non-profit organizations such as universities have different responsibilities and goals than profit-making institutions (Mansbridge, 1998). In order to minimize conflicts and maximize the potential for complementary efforts in collaborative work, Lacy et al. recommend closer monitoring of the nature, goals and outcomes of these relationships; stronger and more creative policies and practices to enhance university interactions with the private sector while protecting the autonomy and freedom of academic scientists; and adequate public agricultural research funding.

Data Analysis of GM Research in Australia

We use publicly available data to elucidate general patterns over the past 15 years in GM research in Australia, in order to address questions about the dominance of private funding in GM research. Australia's OGTR's public website provides a constantly updated table summarizing all applications and authorizations (licenses granted) for intentional release of GMOs into the environment, whether controlled (e.g. as part of a field trial) or released. Although the majority of applications and licenses have been for agricultural and horticultural crops, applications for viruses and vaccines are also listed in the OGTR's table (though excluded here as we focus on crops). While numerous organizations may have been involved in the R&D processes leading to the point of making an application, only the name of the license holder conducting the GMO dealings typically is supplied to the OGTR.

Our analysis of the GM research landscape focuses on key factors that have been discussed in the existing scholarship: the type of organization holding the license, the species or crop, and the trait(s) modified. From the time that OGTR licensing began in Australia in 2002 until 2017, 124 authorizations for Dealings involving Intentional Release (DIRs) for plants used in agriculture and floriculture have been granted to 25 different licensees. Applications for DIRs withdrawn prior to OGTR assessment or still pending and applications for vaccines are excluded from our analysis. Table 4.1, which depicts the distribution of applications for DIRs by entity, shows that the division of private and public licensees is roughly equal. 'Private' entities, including subsidiaries of international corporations such as Monsanto Australia Ltd, small, home-grown companies, university spin-offs and industry-owned

Table 4.1. Applications for DIRs by entity ('other public' and 'other private' are groupings with only one application per entity). (Ankeny, Bray and McKinley, using data from the Office of the Gene Technology Regulator http://www.ogtr.gov.au/internet/ogtr/publishing.nsf/Content/ir-1)

Public		Private	
CSIRO	30	Monsanto Australia	22
Vic DPI/DEDJTR	8	Bayer CropScience	13
University of Queensland	7	Sugar Research Australia/BSES	5
Qld University of Technology	6	Florigene	5
University of Adelaide	4	Dow AgroSciences	3
WA Dept Ag	3	Hexima	3
Other 'public'	3	Pioneer Hi-Bred Australia	2
		Grain Biotech Australia Pty Ltd	2
		Nuseed Pty Ltd	2
		Other 'private'	6
Total	61	Total	63

companies funded by statutory levies (e.g. Sugar Research Australia Ltd) hold or have held 63 out of 124 licenses (51%). 'Public' entities, including universities and federal and state government agencies and research bodies, hold or have held the remaining 61 out of 124 licenses (49%).

Eighty-eight licenses have been granted for dealings with what are considered globally as major crops: cotton, canola, wheat (sometimes combined with barley in a license), rice and maize (although rice and maize are not considered major crops in Australia). Of these licenses, 49 are currently or have been granted to private entities. As shown in Table 4.2, two are for wheat, 14 for canola, and the overwhelming majority for cotton (33). Public-sector licenses for the major crops differ in their distribution: of 39 granted, 19 are for wheat and wheat/barley together, 17 for cotton and one each for canola, maize and rice. Overall, cotton licenses far outnumber wheat/barley and canola combined, with a total of 50 public and private granted, constituting the greatest proportion of DIR license applications. One company, Monsanto Australia Ltd, has held 22 licenses for either cotton or canola. The only organization which has held a greater number of licenses during the period of study is CSIRO, with 30. CSIRO's licenses include 15 for cotton, but also cover a range of other crops such as wheat/barley, grapevine, poppy, maize, rice and safflower.

Comparison of these results with the breakdown for minor crops, which include 'orphan' crops such as cassava, some legumes and coarse varieties of millet, and which are defined by Bender (2013) as those consumed by poorer populations for which there is little commercial market, shows a strong contrast: 17 different 'minor' crops are represented, including sugarcane, safflower, rice, lupin and carnations. Of the 36 licenses granted for these crops, private entities hold 14, but for only six types of crops: three types of flowers modified for qualities such as color (six licenses), sugarcane (five licenses), Indian mustard (two licenses, in addition to the two applications combined with canola) and safflower developed for industrial (i.e. non-food) purposes (one license). Of the remaining minor crops, public entities have applied for 22 licenses for 13 crop and pasture species. These data provide evidence for the patterns observed elsewhere: private, commercial entities unsurprisingly tend to focus on commodities with significant potential for profit. Nevertheless, it also reveals that diverse crops with different purposes and of importance in diverse regions of Australia are represented in GM licensing.

Regarding the claim that private-sector resources tend to be concentrated on plant protection traits, particularly insect resistance (IR) and herbicide tolerance (HT), licensing data show this pattern also tends to hold in Australia, as shown in Table 4.3. Many licenses include multiple traits, some containing as many as six, and the majority containing two or three. The IR and HT traits are by far the most common inclusions in licenses held by the private commercial entities, which far outnumber those held by the public entities for the same traits; 36 of 42 (86%) DIRs which include HT and 27 of 35 (77%) for IR are (or have been) privately held. The picture differs when considering other types of traits, including modifications related to yield, abiotic stress tolerance (e.g. drought or salinity tolerance, extremely important traits in the Australian context), human food composition (e.g. nutrition-related traits), animal nutrition and disease resistance; licenses including these traits are held mostly by public entities.

In summary, this analysis of OGTR publicly available data shows active involvement

Table 4.2. Major and minor crops in DIRs from public and private entities. (Ankeny, Bray and McKinley, using data from the Office of the Gene Technology Regulator http://www.ogtr.gov.au/internet/ogtr/publishing.nsf/Content/ir-1)

Public		Private	
Cotton	17	Cotton	33
Wheat	10	Canola[a]	14
Wheat and barley	9	Wheat	2
Canola	1		
Rice	1		
Maize	1		
Total 'major crops'	39	*Total 'major crops'*	49
Banana	5	Sugarcane	5
Sugarcane	4	Indian mustard (only)	2
Pineapple	2	Carnation	2
White clover	2	Rose	2
Papaya	1	Torenia (bluewings flower)	2
Oilseed poppy	1	Safflower	1
Poppy	1		
Narrow-leafed lupin	1		
Perennial rye and tall fescue grass	1		
Grapevine	1		
Potato	1		
Safflower	1		
Sorghum	1		
Total 'minor crops'	22	*Total 'minor crops'*	14
Grand Total	**61**	**Grand Total**	**63**

[a]Two of the private canola applications include Indian mustard.

Table 4.3. Traits in DIRs from public and private entities. If a GMO contains more than one trait, it is listed in all relevant categories; selectable markers or reporter genes, promoters, and so on are not categorized here. (Ankeny, Bray and McKinley, using data from the Office of the Gene Technology Regulator http://www.ogtr.gov.au/internet/ogtr/publishing.nsf/Content/ir-1)

Public		Private	
Yield	14	Herbicide tolerance	36
Abiotic stress tolerance	13	Insect resistance	27
Composition – food (human nutrition)	13	Abiotic stress tolerance	6
Disease resistance	8	Hybrid breeding system	6
Insect resistance	8	Modified color	5
Herbicide tolerance	6	Plant development	4
Composition – food (processing)	5	Composition – animal nutrition	2
Product quality – food	5	Composition – food (human nutrition)	2
Composition – non-food	4	Composition – non-food	2
Plant development	4	Disease resistance	2
Composition – animal nutrition	2	Yield	2
Product quality – non-food	1	Bioremediation	1
Total	**83**	**Total**	**95**

of both the public and private sectors. Unlike the US situation (e.g. Welsh and Glenna, 2006), private entities arguably do not 'dominate' GM R&D in Australia, given the 49% public/51% private split. The general patterns of focus on certain types of crops and traits in DIRs granted to private versus public entities does appear to parallel those found elsewhere. Although commercially important crops and traits have been prominent especially for private entities, we find evidence of minor and 'orphan' crops and inclusion of non-plant protection traits in the public projects; however, based solely on the quantitative assessment above, it is clear that staple food crops in fact have been relatively neglected.

However, this type of data obscures a range of complexities that must be considered when analyzing patterns in GM research. Some research entities are decidedly hybrid, since they are considered as commercial/private entities, but they rely on statutory levies and hence have certain accountabilities to their stakeholders that mirror responsibilities held by public entities. Most importantly, the publicly available data do not show the extent and nature of research partnerships and collaborations among and between the public and private sectors since (as noted above) only the name of the license holder conducting the GMO dealings typically is supplied to the OGTR. Additional organizations or companies, which may have been involved in the R&D processes leading to the point of making an application, are not available in the public summary data (or often not present explicitly in the licensing documentation). This apparent simplicity of origin obscures the fact that the processes of getting a GMO to where it can be assessed by a regulatory body such as the OGTR is extremely likely to have involved input – financial, technical, material or otherwise – beyond that of the named applicant. We have no reason to assume that who is listed as the applicant is biased in any particular direction, but we note that to perform a full quantitative analysis of the balance and breadth of interests engaged in GM R&D in Australia would require additional source materials, many of which would be difficult to access

given they are likely to be considered commercial-in-confidence.

Hence in the next section we present three brief GM case studies from the Australian context to explore key issues arising and underlying complexities associated with this type of research: (i) drought-tolerant wheat; (ii) high-amylose wheat; and (iii) Vitamin A-enhanced 'super banana'. Not all of these cases were unmitigated successes in commercial terms or with reference to public benefits, but we contend that the mixture of outcomes represented is reflective of typical processes in this domain. We selected these cases because they all involve traits that arguably are associated with the public good in the broad sense, although of course there is also potential for commercial profits in some cases. We also do not intend our choice of these case studies to be construed as endorsement of them, particularly as they have not been without controversy (as we discuss below), but we use them as a springboard for discussion and reflections about future research questions that should be pursued to further illuminate various tensions that exist in this domain.

GM Case Study 1: Drought-Tolerant Wheat

Wheat is the dominant grain crop in Australia and one of its most valuable agricultural exports, second only to beef in 2015 (Xue et al., 2017). According to Wilson et al. (2015), there has been a recent resurgence in GM wheat research, especially in the USA and Australia, following suspension of earlier attempts to commercialize GM wheat due to grower and consumer opposition in Canada and the USA (Eaton, 2011; Kinchy, 2012). In 2004, Monsanto withdrew applications for commercial release of its GM wheat in various countries, including Australia (Schurman and Munro, 2010; ISAAA, 2018).

Among the most common traits targeted for GM are those for tolerance to abiotic stressors such as drought, salinity and frost; ten of the 21 (48%) wheat (or wheat/barley) DIRs issued by the OGTR since 2005 include drought-tolerance traits, with other traits being pursued including enhanced yield, improved nutrient use efficiency, improved grain quality and altered grain composition (OGTR, 2018). The focus on drought tolerance is unsurprising, given that heat and drought stress are having considerable impacts and are considered to be the major challenges to future wheat production in Australia and many other grain-producing regions (Hopkins, 2009; Langridge, 2012). Although breeding for drought tolerance in wheat by conventional methods has been practiced for decades, only 'modest gains' are said to have been achieved thus far (O'Neill, 2010).

In 2006, BASF Plant Science, a plant biotechnology subsidiary of the German chemical company BASF SE, announced an approximately A$28 million investment in a project to develop drought-tolerant wheat in Australia, spanning seven years and involving 25 scientists based at the public- and industry-funded Australian Molecular Plant Breeding Cooperative Research Centre (MPBCRC) (MPBCRC, no date; Hopkins, 2009). First established at the Waite Campus of the University of Adelaide in July 1997 as the Cooperative Research Centre for Molecular Plant Breeding, and intended to benefit the crop and pasture industries, the Centre received a further seven years' funding under the Commonwealth Government's CRC Program in July 2003, continuing in a new location in Victoria as the MPBCRC (which was disbanded in June 2010). A document produced by the MPBCRC (no date) during this period listed six core partners in the CRC (not just the wheat program), including the Victorian Government's Department of Primary Industries (now the Victorian Government Department of Environment and Primary Industries, and hereafter DPI Vic) and ten commercial and industry partners, including BASF, along with its federal government funding.

The gene candidates for the desired trait of drought tolerance were derived from plants (maize and thale cress), a moss and a yeast (Australian Grain, 2007), and were provided by BASF Plant Science (along with genes related to yield increase and resistance to fungal diseases), with MPBCRC providing

'expertise and a patented technique for developing highly effective genetic modifications of wheat' (MPBCRC, no date, p. 9). Rights to commercialize any products resulting from the project were to be held by MPBCRC for Australia, New Zealand and some countries in the developing world, namely those countries assisted by the International Maize and Wheat Improvement Centre (CIMMYT) in Mexico, a non-profit research and training organization with over 500 partners in 100 countries and another core participant in MPBCRC (MPBCRC, no date; CIMMYT, 2016). BASF was to handle commercialization elsewhere (MPBCRC, no date, p. 10).

Although the publicly available information summarized above shows that the research processes prior to application were collaborative efforts between numerous public and private entities under the umbrella of MPBCRC, officially the DIRs were granted by the OGTR to the DPI Vic for the initial GM wheat trials of relevance in 2007 and 2008. DIR 071/2006 permitted trials of GM wheat with drought-tolerance genes, the first such trial in Australia, during the 2007 and 2008 growing season, with DIR 080/2007 covering trials from July 2008 to March 2010 involving new GM lines and continued research on previously approved lines (OGTR, 2016). Following 'very promising' field trials over the two licensing periods, the MPBCRC's then-Chief Executive, Dr Glenn Tong, stressed the need for a cautious approach to interpreting the preliminary results, with many field trials yet to come (Hopkins, 2009).

When MPBCRC ceased operating in June 2010, DPI Vic was expected to continue with the GM wheat program (O'Neill, 2010); whether this research extended beyond the monitoring phase of the trials in question is unknown. Agriculture Victoria Services Pty Ltd – a private company wholly held by the Victorian Government – is responsible for the commercialization and protection (including intellectual property) of novel technologies created by the Agriculture Victoria Research Division, into which DPI Vic was recently rolled. Both licenses were surrendered in March 2016; drought-tolerant GM wheat has yet to be commercialized, with no DIRs pending.

GM Case Study 2:
High-Amylose Wheat

Genetic modification for traits offering potential public health benefits is another area of crop research, which can fall in the realm of 'public good'. This case examines a wheat developed to be high in amylose, a type of resistant starch or functional form of dietary fiber, which is already present in wheat and other whole foods, but which is lacking in typical modern diets (Braidotti, 2016). Increasing the level of resistant starch in wheat, a staple grain, was pursued due to its potential contributions to promoting digestive health, fighting Type 2 diabetes and reducing the risk of bowel cancer, without requiring behavioral changes (Regina *et al.*, 2015; CSIRO, 2017).

High-amylose GM wheat was developed by Arista Cereal Technologies, an Australian–French, public–private joint venture announced in 2006 by the Australian research bodies CSIRO and the Grains Research and Development Corporation (GRDC) and the French-based company Limagrain Céréales Ingrédients (LCI) (CSIRO, 2017). LCI is a subsidiary of the French seed company Limagrain, an international farmer cooperative or grower-owned corporation that 'develops and manufactures authentic and functional cereal ingredients for manufacturers in the food industry' (LCI, no date). Arista was created to allow research, development and commercialization of a high-amylose wheat suitable for processing, utilizing genetic technologies developed by CSIRO's Plant Industry division and Biogemma UK Ltd, a European plant biotechnology company of which Limagrain is a shareholder. Biogemma was founded in 1997 by seed companies and French field crop producers, and shareholders include seed and agricultural finance companies and a French arable crops R&D institute (Biogemma UK Ltd, no date). The GRDC is a statutory corporation founded in 1990 to undertake research, development and extension on behalf of Australian grain growers as well as for the benefit of industry and the public more widely. Primary financial support to the corporation comes from two sources: a grower levy based on the net

farm gate value of the annual production of 25 grain, pulse and oilseed crops, and an Australian government contribution, annually determined and based on the three-year rolling average of the gross value of production of the 25 leviable crops (GRDC, 2018).

A team composed of plant geneticists, agronomists and human nutritionists including researchers affiliated with CSIRO's Food Futures National Research Flagship as well as Biogemma UK Ltd used a combination of GM and conventional breeding techniques to create wheat with the desired qualities (Regina *et al.*, 2006; Braidotti, 2016). It was hoped that the use of some conventional techniques might make the wheat more acceptable to consumers; however, despite early optimism (e.g. Patton, 2006), a clear pathway to regulatory and consumer acceptance of GM wheat has not readily emerged (see Salleh, 2006).

Following successful animal feeding trials, human nutritional studies of the high-amylose wheat were planned. However, activists from environmental organization Greenpeace destroyed the first outdoor trial of the CSIRO GM wheat near Canberra in July 2011; this crop was intended to be used in the first human studies (ABC, 2011). Greenpeace claimed that CSIRO's involvement in the project represented a conflict of interest:

> The web of public-private partnerships that sits behind these research programs is misleading and makes it challenging for the public to know where and to whom their tax dollars are being spent. This lack of transparency makes it difficult for the public to exercise its right to hold the government to account. This difficulty is exacerbated by the secrecy surrounding government documents related to GM plants. Greenpeace's Freedom of Information request for documents about the commercial partnership between CSIRO and Limagrain was refused. The documents are 'commercial in confidence'. Australian taxpayers cannot properly exercise their rights to hold the government to account under these conditions. (2011, n.p.)

Arista owns the intellectual property associated with this project and issued the first license for the high-amylose trait to an American milling company in 2016, with LCI as their partner in breeding the trait into locally adapted wheat varieties (Bay State Milling, 2017). The first crop of enhanced (but non-GM) wheat, grown in Idaho, Oregon and Washington, was harvested in 2017 and was to be milled into trademarked high-fiber wheat flour (CSIRO, 2017). Royalties would be paid per hectare of wheat grown, in principle providing a return on the funding originally allocated through funding of the GRDC by farmer levy and taxpayer funding of the CSIRO (Neales, 2017).

As for the potential future of high-amylose wheat in Australia, it was recently stated that Arista was partnering with a breeding company to develop high-amylose wheat varieties suitable for different regions, and working on ways to produce enough grain for product testing and seeds for initial commercialization. Lindsay Adler from CSIRO and an Arista Director stated that the company was keen to find an Australian licensee who would develop a new product for local and possibly Asian markets (CSIRO, 2017).

GM Case Study 3: Vitamin A-Enhanced Banana

Our third case also concerns the development of a crop with a modification that meets a public health need, a vitamin A-rich banana. One of the world's top-ten food crops by production (Paul *et al.*, 2017), various types of bananas are widely grown in wet tropical and subtropical regions for consuming fresh or cooking, with 85% of production worldwide consumed domestically. In many banana-growing regions, cooking bananas are principal staple foods (particularly in rural areas) as well as subsistence crops; in some regions of Africa and Asia, they also are the major source of dietary starch (Dale *et al.*, 2017a).

Described by lead researcher James Dale as a 'significant humanitarian project' (QUT, 2017), research on the biofortified banana has been underway since at least 2005. Biofortification is the process of increasing the levels of essential nutrients, especially in staple foods, by conventional plant breeding techniques or genetic modification (Bender, 2013; Dale *et al.*, 2017a). Similar to the widely publicized (and highly controversial) GM Golden Rice, the enhanced banana, also orange-fleshed due to its provitamin A (beta carotene) content, is seen as a way to combat vitamin A deficiency, which despite various public health initiatives is still a major problem, particularly in parts of Africa such as Uganda and in Southeast Asia. In addition to the public health aspect of the project, the banana will be engineered to be disease resistant and will be freely available if commercially approved: 'No patents, breeders/variety rights, or commercial rights have been or will be claimed on the pro-vitamin A genes or trait. There will be no technology fees associated with applying the technology in Uganda or elsewhere, nor will there be any additional cost to farmers' (Banana21, 2016, n.p.).

Related GM research has focused on generating resistance to a virulent fungus commonly found in Cavendish bananas (known as Panama disease). Cavendish bananas constitute over 40% of world production and virtually all the export trade in bananas. Virus resistance research has been funded via the ARC Linkage program (LP110100186), in partnership with researchers in the USA, the Netherlands and the Darwin Banana Farming Company, and an Australian banana wholesaler, LaManna (Dale *et al.*, 2017b). This project provides a good example of 'open research', inasmuch as the researchers have communicated about and involved local banana growers in the research, particularly given the economic importance of bananas in Queensland, where the research institute is located and trials were to occur.

Prior to receiving support from the private Bill & Melinda Gates Foundation (BMGF) for the biofortified banana research, researchers at the Centre for Tropical Crops and Biocommodities at Queensland University of Technology (QUT) were working with Ugandan researchers from the National Agricultural Research Organization. They made a successful bid for their Vitamin A banana project in response to a 2004 call from BMGF for expressions of interest regarding solutions for the 'Grand Challenges in Global Health' initiative, including the development of micronutrient-rich staple crops (Fresh Fruit Portal, 2013). In 2005, BMGF began supporting the project, now one of the genetic improvement projects under the umbrella of the 'Banana21' collaboration between scientists at QUT and their Ugandan counterparts, and were joined by the UK Government Department for International Development in 2012 for Phase 3 (Banana21, 2016).

Following initial laboratory work, field trials began in Queensland in 2008 (which also was the first trial of GM bananas in Australia), with QUT obtaining licenses for controlled release. The intention was to transfer the resultant technologies to Uganda for incorporation into the local cultivars and for further field trials (which began there in 2010), and for selection and eventual release there and in other African countries with similar needs (QUT, 2014). However, when approval for human trials in the USA was announced in 2014, controversy arose with protests outside the BMGF headquarters in Seattle and elsewhere. Thus the trials were delayed, not only because of the protests but also because it was found that transporting the fruit was difficult, with no public announcement available about their completion. Uncertainty also has surrounded the process of approval for commercial release of the banana in Uganda: although there now is a regulatory framework for obtaining licenses for GM crop trials, a bill to expand the framework to include GMO release which was approved by Parliament in late 2017 is yet to be signed by the Ugandan president; no GM crops have been approved for commercial sale to date. Therefore, early predictions regarding the release date of the banana have been pushed back as the project continues.

Discussion

While these case studies are admittedly selective, combined with the quantitative analysis presented above they illustrate forms of collaborative research that have largely been overlooked in existing analyses of public–private partnerships, and they reveal different types of funding patterns than conventional collaborative partnerships, involving governmental entities and programs emphasizing both commercial and public benefits. In addition, they encourage us to consider less restricted definitions of 'public benefit'. These points taken together underscore that the potential sources of ethical tensions and conflicts about GM research typically discussed in popular and scholarly literature, notably the undue influence of industrial interests, may need to be reconsidered or at least situated in a much more complex context.

At least in Australia, the partnerships that underlie current GM research are complex webs that, ironically, have frequently been criticized by GM opponents (e.g. by Greenpeace). A key takeaway from our analysis is that many projects involve public entities and private companies and include not only large multinationals but also locally based start-ups and transnational collaborations; the latter examples might not be surprising given that most Australian companies are relatively small (Dodgson et al., 2011). In addition, some of the projects involve non-governmental organizations (NGOs) as well as farmer-associated organizations. Importantly, current analyses of GM research tend to exclude these types of partners, because they have limited presence in the USA, where most previous analysis has occurred, or because they have emerged more recently than the periods covered in previous scholarship. Overlooking projects involving these other types of partners and alignments could indicate an overemphasis on projects aimed at profits and other benefits not primarily aligned with the public good. They also may obscure the fact that some products that might appear to be primarily commercial or economic may contribute to the public good when defined more broadly. For instance, traits that provide environmental benefits also are likely to benefit farmers economically by making crops better adapted to the extreme conditions in Australia. In addition, many projects in this survey were funded not just via university–industry partnerships, but involved governmental funding schemes, such as the CRCs and ARC Linkage programs, as well as NGOs, all of which have mandates to create public goods, but in many cases also use commercial activities to partially fund ongoing and future research.

This analysis suggests the need to reconsider what counts as public benefit. Narrowly construed, 'public benefit' is often taken to be equivalent with promoting global food security, parallel to the values promoted via the Green Revolution, in contrast with the 'gene revolution' associated with GM (e.g. Parayil, 2003; FAO, 2004). However, the range of GM crop types pursued in the Australian research context is broad, with diverse goals within each project. Many of these are well-aligned with public needs and benefits, particularly associated with reducing environment impacts and growth of crops in extreme climatic conditions. It even may be contended that Australia has a distinct role to play in the pursuit of GM research aimed at public benefit given its unique qualities. For instance, Australia is the only 'officially recognized Developed Countr[y]' with a tropical region area (Banana21, 2016), and therefore the only one with a tropical agriculture industry; hence it is well placed to develop various types of bananas that can have benefits in less developed regions and distributed in ways that are not commercialized (the question remains of course about whether these types of initiatives are appropriate in other terms, but this concern is not our primary focus in this chapter). The presence of projects including traits other than those associated with plant protection raises questions about fears that private agendas are swamping out other types of R&D priorities in GM research, which does not seem to be supported by the current Australian data.

Conclusions

It is clear that there are various types of tensions at work surrounding GM agriculture. A particularly important tension involves what role industry and other non-public entities should play in the development of GM technology and how the distribution of benefits should be decided. The examples presented

here from the Australian context do not show or suggest that any or all GM research is good, or worth pursuing or supporting. There are numerous factors that need to be considered to develop a definitive analysis of the impact of GM research to date and future potential (or lack thereof). Instead, our contribution provides empirical information on the current state of play that allows us to widen our dialogue about these issues. We concur with Vallas and Kleinman, who contend that biotechnology (and GM research in particular) is a domain

> marked by an increasing commingling of normative codes and practices from two previously relatively distinct institutional domains, leading to the emergence of a knowledge regime that is fraught with tension, contradiction and inconsistency ... Far from demanding resolution, such tensions can in fact serve as a source of creative dynamism, dialogue and reflexivity, compelling the various parties to justify their assumptions, to engage in dialogue with those whose orientation differs from their own and thus to make possible a deeper and more innovative understanding of the major tasks at hand. (2008, p. 306)

As Vallas and Kleinman note, rather than viewing these complexities and tensions as difficulties, we should treat them as assets particularly because of their ambiguity (see also Stark, 2001).

Our study highlights future research topics to be pursued in order to enrich our understanding of the tensions underlying GM research and domains facing similar issues. First, due to the limitations of the existing quantitative data, more up-to-date, qualitative evidence should be gathered about how priorities are set, including the potential for increasing public engagement and participation in decision making, and how research is structured, especially with regard to access to products, intellectual property arrangements, and so on. Admittedly, this type of approach may be limited due to commercial-in-confidence considerations, but this potential difficulty is not sufficient to warrant ignoring these issues. Second, more exploration is needed of scientists' views on how they fulfil their 'social license' (Raman and Mohr, 2014) and assess and respond to public priorities and needs, particularly in the Australian case given the OGTR's lack of mandate in this regard. Finally, GM research has taken hold in numerous less high-profile locales, and collaborative practices in South America, Canada, Eastern Europe and Asia are worthy of more scholarly analysis than has occurred to date.

Our analysis also shows that ignoring funding arrangements for the development of GM crops to a public–private binary obscures the complex networks and types of arrangements that are typical of contemporary science. It could be argued that the funding arrangements for the development of GMOs are less important to consider than evaluating how the GMO will be deployed within food production systems, neoliberal or otherwise, how these arrangements are regulated and monitored, and who in fact can benefit and how.

Acknowledgments: Research for this paper was in part funded by an Australian Research Council Discovery Project grant (DP1401025477) "Making Plants Better, Making Australia Better? A History of Genetic Modification Science, Policy, and Community Attitudes in Australia." We are grateful to audiences at the Australian Association for History, Philosophy, and Social Studies of Science 2015 conference and University of Notre Dame's 2015 Collaboration Conundrum Conference for feedback on previous versions of this paper.

Disclosure: Dr Heather Bray was Public Engagement Officer with the Waite Research Institute between 2011-14 and between 2003-10. She developed and delivered community and schools education programs on GM crops for the Molecular Plant Breeding CRC.

References

ABC (2011) Greenpeace blasted for GM vandalism. ABC News, 15 July. Available at: http://www.abc.net.au/news/2011-07-15/scientists-condemn-greenpeace-gm-cull/2795482 (accessed 5 January 2018).

Ankeny, R.A. and Bray, H.J. (2016) 'If we're happy to eat it, why wouldn't we be happy to feed it to our children?' Articulating the complexities underlying women's ethical views on genetically modified food. *International Journal on Feminist Approaches to Bioethics* 9, 166-191. doi:10.3138/ijfab.9.1.166

Ankeny, R.A. and Bray, H.J. (2018) Genetically-modified food. In: Barnhill, A., Doggett, T. and Egan, A. (eds) *Oxford Handbook of Food Ethics*, Oxford University Press, Oxford, pp. 95–111.

Australian Academy of Science (1980) *Recombinant DNA: An Australian Perspective*. Australian Academy of Science, Canberra.

Australian Grain (2007) First Australian trial of drought tolerant GM wheat gets approval. *Australian Grain* 17, 17.

Banana21 (2016) Home page. Available at: http://www.banana21.org/index.html (accessed 10 January 2018).

Bay State Milling (2017) Bay State Milling Company launches wheat flour with fiber benefits. Available at: http://www.baystatemilling.com/media/bay-state-milling-company-launches-wheat-flour-with-fiber-benefits/ (accessed 5 January 2018).

Bender, D.A. (2013) *A Dictionary of Food and Nutrition*, 4th edn [online]. Oxford University Press, Oxford. doi:10.1093/acref/9780191752391.001.0001

Biogemma UK Ltd (n.d.) About Biogemma. Available at: http://www.biogemma.com/index.php/about-biogemma (accessed 4 January 2018).

Biscotti, D., Glenna, L.L., Lacy, W.B. and Welsh, R. (2009) The 'independent' investigator: How academic scientists construct their professional identity in university–industry agricultural biotechnology research collaborations. *Economic Sociology of Work, Research in the Sociology of Work* 18, 261–285. doi:10.1108/S0277-2833(2009)0000018013

Blumenthal, D., Causino, N., Campbell, E. and Louis, K.S. (1996) Relationships between academic institutions and industry in the life sciences – an industry survey. *New England Journal of Medicine* 334, 368–373. doi:10.1056/NEJM199602083340606

Bok, D. (2003) *Universities in the Marketplace: The Commercialization of Higher Education*. Princeton University Press, Princeton, New Jersey.

Braidotti, G. (2016) 'Tweaked' wheat starch may open new global health market. *GroundCover* 121. Available at: https://grdc.com.au/resources-and-publications/groundcover/ground-cover-issue-121-mar-apr-2016/tweaked-wheat-starch-may-open-new-global-health-market (accessed 20 December 2017).

Bray, H.J. and Ankeny, R.A. (2015) What do food labels teach people about food ethics? In: Swan, E. and Flowers, R (eds) *Food Pedagogies*, Ashgate, London, pp. 185–200.

Bray, H.J. and Ankeny, R.A. (2017) Not just about 'the science': Science education and attitudes to genetically-modified foods among women in Australia. *New Genetics and Society* 36, 1–21. doi:10.1080/14636778.2017.1287561

Busch, L., Lacy, W.B., Burkhardt, J. and Lacy, L.R. (1991) *Plant, Power and Profit: Social, Economic, and Ethical Consequences of the New Biotechnologies*. Basil Blackwell, Oxford.

CIMMYT (2016) Our work. Available at: http://www.cimmyt.org/our-work/ (accessed 5 January 2018).

Collier, A. (2007) Australian framework for the commercialisation of university scientific research. *Prometheus* 25: 51–68. doi: 10.1080/08109020601172894

Commonwealth of Australia (2001) *Backing Australia's Ability: An Innovation Action Plan for the Future*. Commonwealth Government, Department of Industry, Science and Resources, Canberra.

Commonwealth of Australia (2010) *Australian Innovation System Report 2010*. Australian Government Department of Innovation, Industry, Science and Research, Canberra.

Cousins, Y.L., Lyon, B.R. and Llewellyn, D.J. (1991) Transformation of an Australian cotton cultivar: Prospects for cotton improvement through genetic engineering. *Australian Journal of Plant Physiology* 18, 481–494. doi:10.1071/PP9910481

CSIRO (2017) Wheat: a kick in the guts for fighting diseases. Available at: https://www.csiro.au/en/News/News-releases/2017/Wheat-a-kick-in-the-guts-for-fighting-diseases (accessed 20 December 2017).

Dale, J., Paul, J.-Y., Dugdale, B. and Harding, R. (2017a) Modifying bananas: From transgenics to organics? *Sustainability* 9, 333–346. doi:10.3390/su9030333

Dale, J., *et al.* (2017b) Transgenic Cavendish bananas with resistance to Fusarium wilt tropical race 4. *Nature Communications* 8, 1496. doi:10.1038/s41467-017-01670-6

Davidson, S. (2003) Biotech cotton: a budding field: Steve Davidson sums up the performance of genetically-modified cotton in Australia. *Ecos* 114, 28–31.

Dodgson, M., Hughes, A., Foster, J. and Metcalfe, S. (2011) Systems thinking, market failure, and the development of innovation policy: The case of Australia. *Research Policy* 40, 1145–1156. doi:10.1016/j.respol.2011.05.015

Doering, D.S. (2004) *Designing Genes: Aiming for Safety and Sustainability in U.S. Agriculture and Biotechnology*. World Resources Institute, Washington, DC.

Eaton, E. (2011) Contesting the value(s) of GM wheat on the Canadian prairies. *New Political Economy* 16, 501–521. doi:10.1080/13563467.2011.519021

Elliott, C. (2010) *White Coat Black Hat: Adventures on the Dark Side of Medicine*. Beacon Press, New York.

Encyclopedia of Australian Science (2011) Cooperative Research Centres Program (1990–). Available at: http://www.eoas.info/biogs/A001898b.htm (accessed 20 December 2017).

Ervin, D., *et al.* (2001) *Transgenic Crops: An Environmental Assessment*. Policy Studies Report, Henry A. Wallace Center for Agricultural and Environmental Policy at Winrock International, Arlington, Virginia.

Etzkowitz, H. and Webster, A. (1998) Entrepreneurial science: The second academic revolution. In: Etzkowitz, H., Webster, A. and Healey, P. (eds) *Capitalizing Knowledge: New Intersections of Industry and Academia*, SUNY Press, Albany, New York, pp. 21–46.

FAO (2004) *The State of Food and Agriculture. Agricultural Biotechnology: Meeting the Needs of the Poor?* Available at: http://www.fao.org/docrep/006/Y5160E/Y5160E00.htm (accessed 15 January 2018).

Fresh Fruit Portal (2013) GM bananas: From nutrition to disease resistance. Available at: https://www.fresh-fruitportal.com/news/2013/08/23/gm-bananas-from-nutrition-to-disease-resistance/?country=others (accessed 6 January 2018).

Gibbons, M., *et al.* (1994) *The New Production of Knowledge: The Dynamics of Science and Research in Contemporary Societies*. Sage, Thousand Oaks, California.

GMAC (1996) Annual report 1995–1996. Commonwealth of Australia, Canberra.

GMAC (1997) Annual report 1996–1997. Commonwealth of Australia, Canberra.

GRDC (2018) What we do. Available at: https://grdc.com.au/about/what-we-do (accessed 21 December 2017).

Greenpeace (2011) Q & A on GM wheat trial action. Available at: http://www.greenpeace.org/australia/en/what-we-do/Food/resources/FAQs/Q--A-on-GM-Wheat-trial-action/ (accessed 4 January 2018).

Gregory, R. (1993) The Australian innovation system. In: Nelson, R. (ed.) *National Innovation Systems: A Comparative Analysis*. Oxford University Press, New York, pp. 324–352.

Hackett, E.J. (2005) Essential tensions: Identity, control, and risk in research. *Social Studies of Science* 35, 787–826. doi:10.1177/0306312705056045

Hain, M., Cocklin, C. and Gibbs, D. (2002) Regulating biosciences: The Gene Technology Act 2000. *Environmental and Planning Law Journal* 19, 163–179.

Hindmarsh, R. (2008) *Edging Towards BioUtopia: A New Politics of Reordering Life and the Democratic Challenge*. University of Western Australia Press, Crawley, Australia.

Hopkins, P. (2009) Field trials of GM projects yield 'promising' results. *The Age*, Melbourne, Australia, 30 March, p. 2.

ISAAA (2016) ISAAA Briefs, Brief 52, Global status of commercialized biotech/GM crops: 2016. Available at: http://www.isaaa.org/search/results.asp?cx=005763828013670756947%3Ag9mmea06vvc&cof=FORID%3A10&ie=UTF-8&q=global+status&sa=Go (accessed 23 January 2018).

ISAAA (2018) GM approval database, GM crop events list, Event Name: MON71800. Available at: www.isaaa.org/gmapprovaldatabase/event/default.asp?EventID=237 (accessed 26 January 2018).

Kerr, A. (2011) GM crops: A mini-review. *Australasian Plant Pathology* 40, 449–452. doi:10.1007/s13313-011-0073-7

Kinchy, A.J. (2012) *Seeds, Science, and Struggle: The Global Politics of Transgenic Crops*. MIT Press, Cambridge, Massachusetts.

Kleinman, D.L. (2010) The commercialization of academic culture and the future of the university. In: Radder, H. (ed.) *The Commodification of Academic Research: Science and the Modern University*. University of Pittsburgh Press, Pittsburgh, Pennsylvania, pp. 24–43.

Krimsky, S. (2003) *Science in the Private Interest: Has the Lure of Profits Corrupted Biomedical Research?* Rowan & Littlefield Publishers, Oxford.

Krimsky, S., Campbell, E.G. and Blumenthal, D. (1999) Perils of university-industry collaboration. *Issues in Science and Technology* 16, 14–15.

Lacy, W.B., Glenna, L.L., Biscotti, D., Welsh, R. and Clancy, K. (2014) The two cultures of science: Implications for university–industry relationships in the U.S. agriculture biotechnology. *Journal of Integrative Agriculture* 13, 455–466. doi:10.1016/S2095-3119(13)60667-X

Langridge, P. (2012) Genomics: Decoding our daily bread. *Nature* 491, 678–680. doi:10.1038/491678a

Lawson, C. (2002) Risk assessment in the regulation of gene technology under the Gene Technology Act 2000 (Cth) and the Gene Technology Regulations 2001 (Cth). *Environmental Planning and Law Journal* 19, 195–216.

LCI (Limagrain Céréales Ingrédients) (n.d.) Homepage. Available at: http://www.lci.limagrain.com/index-en (accessed 19 December 2017).

Levidow, L. and Carr, S. (2000) Unsound science? Transatlantic regulatory disputes over GM crops. *International Journal of Biotechnology* 2, 257–273. doi:10.1504/IJBT.2000.000131

Lockie, S., Lawrence, G., Lyons, K. and Grice, J. (2005) Factors underlying support or opposition to biotechnology among Australian food consumers and implications for retailer-led food regulation. *Food Policy* 30, 399–418. doi:10.1016/j.foodpol.2005.06.001

Lu, C.-Y., Nugent, G., Wardley-Richardson, T., Chandler, S.F., Young, R., *et al.* (1991) Agrobacterium-mediated transformation of a carnation (*Dianthus caryophyllus* L.). *Bio/Technology* 9, 864–868. doi:10.1038/nbt0991-864

Mansbridge, J. (1998) On the contested nature of the public good. In: Powell, W.W. and Clemens, E.S. (eds) *Private Action and the Public Good*. Yale University Press, New Haven, Connecticut, pp. 3–19.

Mowery, D.C., Nelson, R.R., Sampat, B. and Ziedonis, A.A. (2004) *Ivory Tower and Industrial Innovation: University–Industry Technology Transfer Before and After the Bayh-Dole Act*. Stanford University Press, Stanford, California.

MPBCRC (n.d.) Submission. Review of the National Innovation System – Cooperative Research Centres Programme. Available at: https://dfat.gov.au/trade/topics/review-of-export-policies-programs/Documents/VictorianAgriBiosciencesCentre.pdf (accessed 20 December 2017).

Neales, S. (2013) An inconvenient truth. *The Australian*, 18 January. Available at: http://www.theaustralian.com.au/news/features/an-inconvenient-truth/story-e6frg6z6-1226556153378 (accessed 25 January 2018).

Neales, S. (2017) CSIRO super wheat great for guts. *The Australian*, Canberra, Australia, 15 December, p. 27.

OGTR (2016) Licence for Dealings involving an Intentional Release of a GMO into the environment, Licence No.: DIR 080/2007. Available at: http://search.health.gov.au/s/search.html?query=surrendered+licence+DIR+080%2F2007&collection=health&profile=ogtr&Submit (accessed 5 January 2018).

OGTR (2018) Table of applications and authorisations for Dealings involving Intentional Release (DIR) into the environment. Available at: http://www.ogtr.gov.au/internet/ogtr/publishing.nsf/Content/ir-1 (accessed 4 January 2018).

O'Neill, G. (2010) Biotech pathway cleared for stepchange yield gains. *GroundCover* 87. Available at: https://grdc.com.au/resources-and-publications/groundcover/ground-cover-issue-87-july-august-2010/biotech-pathway-cleared-for-stepchange-yield-gains (accessed 20 December 2017).

Paarlberg, R.L. (2000) *Governing the GM Crop Revolution: Policy Choices for Developing Countries*. Food, Agriculture and the Environment Discussion Paper 33. IFPRI, Washington, DC.

Parayil, G. (2003) Mapping technological trajectories of the Green Revolution and the Gene Revolution from modernization to globalization.

Research Policy 32, 971–990. doi:10.1016/
S0048-7333(02)00106-3

Patton, D. (2006) Limagrain, GRDC to bring GM healthy wheat to market. *Food Navigator.* Available at: https://www.foodnavigator.com/Article/2006/11/02/Limagrain-GRDC-to-bring-GM-healthy-wheat-to-market (accessed 20 December 2017).

Paul, J.-Y., *et al.* (2017) Golden bananas in the field: Elevated fruit pro-vitamin A from the expression of a single banana transgene. *Plant Biotechnology Journal* 15, 520–532. doi:10.1111/pbi.12650

QUT (2014) Super bananas-world first human trial. Available at: https://www.qut.edu.au/news?news-id=74075 (accessed 21 December 2017).

QUT (2017) QUT develops golden bananas high in vitamin A. Available at: https://www.qut.edu.au/news?news-id=119796 (accessed 21 December 2017).

Radder, H. (ed.) (2010) *The Commodification of Academic Research: Science and the Modern University.* University of Pittsburgh Press, Pittsburgh, Pennsylvania.

Raman, S. and Mohr, A. (2014) A social licence for science: Capturing the public or co-constructing research? *Social Epistemology* 28, 258–276. doi:10.1080/02691728.2014.922642

Regina, A., *et al.* (2006) High-amylose wheat generated by RNA interference improves indices of large-bowel health in rats. *Proceedings of the National Academy of Sciences of the United States of America* 103, 3546–3551. doi:10.1073/pnas.0510737103

Regina, A., *et al.* (2015) A genetic strategy generating wheat with very high amylose content. *Plant Biotechnology Journal* 13, 1276–1286. doi:10.1111/pbi.12345

Salleh, A. (2006) Gene silencing yields high-fibre wheat. *ABC Science,* 28 February. Available at: http://www.abc.net.au/science/articles/2006/02/28/1576926.htm (accessed 21 December 2017).

Schurman, R. and Munro, W.A. (2010) *Fighting for the Future of Food: Activists Versus Agribusiness in the Struggle over Technology.* University of Minnesota Press, Minneapolis, Minnesota.

Sismondo, S. (2008) How pharmaceutical industry funding affects trial outcomes: Causal structures and responses. *Social Science & Medicine* 66, 1909–1914.doi:10.1016/j.socscimed.2008.01.010

Slaughter, S. and Leslie, L. (1997) *Academic Capitalism: Politics, Policies and the Entrepreneurial University.* Johns Hopkins University Press, Baltimore, Maryland.

Stark, D. (2001) Ambiguous assets for uncertain environments: Heterarchy in postsocialist firms. In: DiMaggio, P. (ed.) *The Twentieth Century Firm: Changing Economic Organization in International Perspective.* Princeton University Press, Princeton, New Jersey, pp. 69–104.

Thompson, P.B. (2007) *Food Biotechnology in Ethical Perspective,* 2nd edn. Springer, Dordrecht, The Netherlands.

Tribe, D. (2012) Gene technology regulation in Australia: A decade of a federal implementation of a statutory legal code in a context of constituent states taking divergent positions. *GM Crops & Food* 3, 21–29. doi:10.4161/gmcr.18606

Vallas, S.P. and Kleinman, D.L. (2008) Contradiction, convergence and the knowledge economy: The confluence of academic and commercial biotechnology. *Socio-Economic Review* 6, 283–311. doi:10.1093/ser/mwl035

Welsh, R. and Glenna, L.L. (2006) Considering the role of the university in conducting research on agri-biotechnologies. *Social Studies of Science* 36, 929–942. doi:10.1177/0306312706060062

Wickson, F. (2007) From risk to uncertainty in the regulation of GMOs: Social theory and Australian practice. *New Genetics and Society* 26, 325–339. doi:10.1080/14636770701701832

Wilson, W., Shakya, S. and Dahl, B. (2015) Valuing new random genetically modified (GM) traits with real options: The case of drought-tolerant wheat. *Agricultural Finance Review* 75, 213–229. doi:10.1108/AFR-05-2014-0014

Xue, Q., Rudd, J., Bell, J., Marek, T. and Liu, S. (2017) Improving water management in winter wheat. In: Langridge, P. (ed.) *Achieving Sustainable Cultivation of Wheat, Volume 2: Cultivation Techniques.* Burleigh Dodds Science Publishing, Cambridge, pp. 63–84.

5 Three Models of Public Opinion and Public Interest for Agricultural Biotechnology: Precautionary, Conventional and Accommodative

Duane Windsor*

Jones Graduate School of Business, Rice University, Houston, Texas, USA

Introduction

The normative research question addressed in this chapter is whether the public interest means agricultural biotechnology should lead (and thus reshape) public opinion or wait for public opinion to become sufficiently supportive. There is no single overall answer to the normative question, because agricultural biotechnology controversially includes genetically modified organisms (GMOs) for human consumption and medical applications. Public opinion and public interest legitimately vary by country, while pro-GMO science remains disputable. The public interest presently depends on the value model in a particular country. This chapter presents three value models that combine public opinion and public interest differently. These three models form a rough continuum from precautionary through conventional to accommodative. The European Union (EU) precautionary model is anti-GMO. Public opinion leads and biotechnology follows. The conventional model, associated with some agriculturally oriented countries, is pro-domestic rather than anti-GMO. Here also public opinion leads. The US accommodative model is more GMO-tolerant than pro-GMO. Agricultural biotechnology leads. There is presently no ethical standard for selecting among these three value models.

The three-model framework reveals a profound, unresolved ethical tension within each model. Scientists and GMO-designer businesses are pro-GMO. They can advance biotechnology faster than public opinion can adjust. Science and business can act first and should attempt to shape public opinion and obtain government licensing; or science and business can and should wait for public opinion. Government lies between business (lobbying) and public opinion (citizen consumers and voters). The three models make clear that the ethical problems embedded in GMO production and consumption can be resolved in different ways. Accommodation, conventionalism and precaution are different ethical positions. This assessment could change with future information – concerning in part hunger and poverty alleviation and medical applications. The answer to the normative question thus varies by model. The risk of the precautionary and conventional models is that burdens are imposed on consumers through insufficient access to GMO products. The risk of the accommodative model is that science proves to be wrong, at least in details.

Public opinion, public interest and ethical tension interact as follows. If GMOs markedly

* E-mail: odw@rice.edu

improve human welfare – which is in the public interest – then it is ethical for businesses to lead with strong expert support and arguably unethical to oppose GMOs on the basis of purely local interests. Public opinion should follow and can be influenced. If GMOs do not improve human welfare, then GMOs are simply a substitute for conventional products and pro-GMO business leadership is purely self-interest. The public interest is not in GMOs and public opinion can lead. If GMOs markedly reduce human welfare, then pro-GMO leadership is clearly unethical. Even if GMOs are perfectly safe but do not markedly improve human welfare, there is a public interest choice for each country to make between GMO and conventional agricultural approaches. This choice can reflect public opinion.

Growth in development, cultivation and consumption of GMO foods has been dramatic in some areas of the world while strongly resisted in others. Some major multinational corporations, international organizations and anti-GMO activists have influenced this pattern of growth. Individuals and countries have choices concerning agricultural biotechnology; these choices are shaped by varying public opinion intertwined with varying conceptions of public interest. The evidence in favor of positive human welfare effects is not yet sufficiently strong to invalidate the precautionary or conventional models. In these circumstances, individuals and countries can choose among the three models – as empirically they tend to do presently. Trading arrangements – within the World Trade Organization (WTO) dispute-resolution process – tend to foster import of GMOs from accommodative countries into precautionary and conventional countries without compelling those countries to undertake GMO production. The models also assist in bridging gaps among public opinion, business interest and expert advice. The bridge is different in each model. New technology development leads in the accommodative model; public opinion leads in the precautionary model. Conventional model public opinion is shaped by business interest (especially domestic agriculture) balancing export opportunities and import competition.

The 1992 Convention on Biological Diversity (1992, n.p.) defines biotechnology broadly as 'biological systems, living organisms, or derivatives thereof, to make or modify products for specific use'. Aspects of agricultural biotechnology are controversial and generative of difficult ethical tensions among interests, values and rights. An interest is an economic stake: businesses, including agricultural, and consumers have economic stakes. A value is what an individual prizes. Anti-GMO individuals value something else in opposing GMOs. A right is what an individual claims: anti-GMO individuals claim a right to non-GMO products; other individuals may claim a right to improved medical care including genetic engineering.

Public opinion and public interest conceptions are intertwined, varying by country. Public opinion is the distribution of multifaceted views among individuals about risk, precaution, business and science – and thus about public interest. A consumer may be concerned with risk, and a business with profit; both may be concerned with the public policy framework affecting risk and profit. One key aspect of the controversy over GMOs is that food is consumed with long-term health effects difficult to predict with confidence; another key aspect is that new biotechnology might escape into the natural environment or alter the composition of natural populations.

The public interest may take substantive definition from context. Increasing rice production to feed growing populations more efficiently and effectively is different from improving the profit margins of agricultural biotechnology firms. Public interest is the general welfare of the commonwealth. In the 1960s, there was a scholarly debate over whether this vague idea should and could be defined. There remain two opposing conceptions of public interest. The first conception involves comparison of expected benefits to expected costs of a change: rational beneficiaries and their elected representatives should vote for the choice (Livermore, 2014). This conception is consistent with prescriptive scientific judgment on GMOs in an accommodative context. The second conception is a direct aggregation of private interests.

There is no public interest as such: there may simply be competing interest groups. There may be incompatible conceptions of morality and social interest (Baumol, 1976/2016). In environmental, safety and health policies, there may be a case to be made that ethical standards are superior to cost–benefit decisions and that non-marketed benefits and costs should not be monetized (Kelman, 1981). This conception is consistent with public opinion in precautionary and conventional contexts.

Some advocates attribute the EU's precautionary approach to political and non-scientific considerations (Levidow and Carr, 2000; Renn, 2007) and regard the US approach as scientific and non-political in contrast. Both of these 'descriptions' are partial caricatures. Levidow and Carr point out that the US approach involves 'value-laden features of safety claims, … weak scientific basis, … normative framing and … socio-political influences' (2000, p. 257).

Herring and Paarlberg (2016) make the following points. Green Revolution crop technologies were readily accepted, in contrast to genetic engineering. The opposition to the latter concerns mostly agriculture, whereas medical and pharmaceutical applications have been more readily accepted. The acceptance of agricultural biotechnology varies considerably by country. These authors argue that outcomes cannot be attributed simply to summation of weighted material interests. Rather there is a social construction of risk understanding, politics and social psychology. In addition to the classic interest group and regulator explanation, there has been a transnational diffusion of pro-GMO and anti-GMO ideologies. The economic effects fall on poorer farmers and poorer countries.

Public Opinion, Science and Business

The GMO debate is one of the most controversial science issues of the 21st century (Kangmennaang et al., 2016). Panchin and Tuzhikov (2017) argue that public perception of GMOs is overall negative, and that this negative perception retards GMO crop development and commercialization in developed and developing countries.

Public opinion

Public opinion on GMOs is a compound of different considerations (Bukenya and Wright, 2007; Malyska et al., 2016). One is consumer willingness to bear risk in eating GMO crops. A second is confidence in scientific consensus that GMOs are safe to eat. A third is concern about escape of GMO crops and contamination of non-GMO crops. A fourth, particularly for farmers, is the cost of GMO seeds in relationship to the cost of pesticides. A fifth is preference for non-GMO including organic and local crops. A sixth is a generalized concern about change in the environment. Accordingly, individual views range from anti-GMO (especially clustered in the EU) to GMO-tolerance (especially clustered in the USA).

A 2015 Pew Center survey reported that two-thirds of more than 2000 American respondents did not agree that scientists fully understand potential health effects of GMOs; and those respondents are the ones concerned for personal safety in consumption of GMO foods (Funk and Rainie, 2015). This negative view did not vary with political orientation, level of education or standard demographic characteristics and was reported even by two-thirds of respondents with a four-year college degree in a science field (Ramsey and Brodwin, 2015).

Scientific arguments do not satisfy public concerns over the combination of health and environmental risks. A laboratory experiment with Brazilian consumers provided presence or absence of genetic modification in combination with product benefits (improving environment, price and shelf life). Genetic modification in a product generated negative evaluations of the business and intention to avoid the product, not offset by product benefits. Consumers reporting high involvement with GMO foods did present more favorable product intention. Demographic variables, including gender and age, influenced attitudes in the experiment (Matos, 2004).

Concern for cross-contamination from escaped GMO crops is not trivial (Robaey, 2016). Increasing attention to GMO screening is a social cost to avoid contamination (Singh et al., 2016). Genetic pollution is likely to prove costly to farmers growing conventional (including organic) crops, resulting in legal issues of both damage and patent infringement (Heald and Smith, 2006). Genetic pollution could prove irreversible. Genetic engineering may both enhance and threaten biodiversity (Weih and Polle, 2016). Genetic engineering could help endangered species, control unwanted species and permit de-extinction (Corlett, 2017). These multiple options make assessment of genetic engineering difficult.

Genetic engineering is not a speeded-up version of traditional trial-and-error cross-breeding hybridization of long-established species: it artificially manipulates genetic material in a laboratory for introduction into another species (Fairfield-Sonn, 2016). GMOs are such a highly miscellaneous set of products (including animals, plants and microorganisms) arising from multiple biotechnology methods that the label itself 'has very poor semantic value' (Tagliabue, 2016a, p. 1). If so, then neither blanket anti-GMO nor blanket GMO safety positions are reasonable: scientific consensus involves unique case-by-case assessment of each individual biotechnology invention's safety (Tagliabue, 2016a). Similar problems may arise with biopharming for medical and industrial applications which do not suffer directly from association with human consumption (Bratspies, 2004).

Scientific consensus

There is strong scientific consensus in favor of the safety of GMOs for human consumption (Fahlgren et al., 2016). The explicit consensus includes the American Association for the Advancement of Science (2012), the (US) National Academies of Sciences, Engineering, and Medicine (2016) and the European Food Safety Authority (2013). Reviews of the scientific literature report evidence in support of safety. Nicolia et al. (2014) evaluated 1783 documents (published 2002–2012) concerning, favorably, multiple aspects of GMO crop safety. Panchin and Tuzhikov (2017) conclude that certain negative studies are flawed in statistical analysis of data, and corrected for these flaws do not provide significant evidence of GMO harms.

Nearly 40% (at least 107 of 296) of living Nobel laureates in an open letter criticized Greenpeace for its opposition to GMOs, with specific focus on Golden Rice (Lyon, 2016). Golden Rice (the term means that the rice contains beta carotene) might help relieve vitamin A deficiency in developing countries, which is responsible for more than a million deaths and 250,000 to 500,000 cases of childhood blindness annually, according to the open letter (Lyon, 2016). Greenpeace has argued that GMOs involve 'genetic pollution' of the world (Greenpeace, 2016). Greenpeace responded to the Nobel open letter that Golden Rice has failed and is not currently for sale, and has not been proven to address vitamin A deficiency.

Business developments

Important merger and acquisition activity in the global seed market aims at controlling valuable intellectual property rights and markets (Carbonell, 2016). The result had been six important global conglomerates: BASF and Bayer (Germany), Dow, DuPont and Monsanto (USA) and Syngenta (Switzerland). US and EU regulators delayed a US$130 billion Dow DuPont merger over anti-trust issues (DowDuPont, 2016; Reuters, 2016a), which was finally completed in September 2017. The EU stance concerning GMOs may be affected by a potential Bayer–Monsanto combination likely completing in early 2018. The resulting company would control more than 25% of the world market for seeds and pesticides. The transaction would combine Bayer's crop chemicals business (the second largest in the world after Syngenta) with Monsanto's seeds business (the largest in the world). In 2015, Monsanto had tried to acquire Syngenta, which was subsequently acquired by the China National Chemical Corp. (ChemChina), a Chinese state-owned enterprise. (Information in this paragraph comes from Reuters, 2016b.) If the Bayer–Monsanto deal goes through, three companies would control about 70% of the global pesticide market and as much as 80% of the US corn-seed market; farmers would face two large seed producers (Bayer–Monsanto and DowDuPont) and two large chemical producers (Bayer–Monsanto and Syngenta) (Bartz and Roumeliotis, 2016).

Three Alternative Models of Public Opinion and Public Interest

The precautionary model is most characteristic of the EU. The conventional model is most characteristic of certain other countries with important agricultural industries: Argentina, Brazil, Japan and New Zealand are instances. The accommodative model is most characteristic of the USA.

Precautionary model

The precautionary model is hostile to agricultural biotechnology, and especially GMOs. Much of this technology and associated products originates (or is portrayed as originating) in the USA. Principle 7 of the UN Global Compact explicitly states the precautionary principle concerning the natural environment as follows. An attribute of precaution is that proof of human safety and environmental protection must be extremely strong (or beyond reasonable doubt): doubt can always be expressed. This precaution may well overlap with elements of conventional production and green production/consumption. In this model, the public interest reflects precaution and green orientation (European Commission, 2015); in addition, domestic business is protected from what amounts to foreign (US) competition through

innovation. The precautionary principle is highly non-specific: it is a general stance against GMOs and evidence for introducing GMOs effectively must approach 100% certainty.

Under WTO pressure, the European Commission has been approving specific GMOs for import only as food and feed and various countries have opted out of GMO approval (Perez 2006; Ferer, 2016; Tagliabue, 2016b). The WTO does have rules that permit import restriction on health or life concerns, but these restrictions depend on scientific information (Broude, 2006). The WTO includes an Agreement on the Application of Sanitary and Phytosanitary Standards. In May 2003, the WTO received a case against the EU from the USA, Canada and Argentina; in November 2006, the WTO ruled against the EU precautionary approach (Disdier and Fontagne, 2009). The EU barred GMO imports during the period 1998–2004. A study suggested that the moratorium and product-specific measures by the EU and safeguard measures by Germany, Greece and Italy did reduce imports (Sindico, 2005). The EU now basically resists GMO crop agriculture, while permitting GMO imports for consumption – with labeling – under WTO pressure. The EU approach includes considerable scope for public discussion. The EU also has moved in the direction of greater local option (Bodiguel, 2016).

An extension of the EU precautionary approach is the effort of businesses to undertake voluntary non-GMO labeling and advertising as a strategic advantage with consumers. Voluntary labeling and promotion may involve possible legal liability and adverse consumer reaction if there should prove to be GMO contamination (Wallace, 2016). Ghozzi et al. (2016) provide a case study of the non-GMO approach in the poultry industry of France and Italy, based on interviews with main actors at five stages of the industry chain (from animal feed and crop production to the final retailer). A study compared GMO-free foods and GMO foods using a proprietary database containing detailed product descriptions at the retail level (Goodwin et al., 2016). In an item to item comparison, GMO-free food costs on average

33% more. On a per-ounce basis, the average difference was 73% more. For a typical market basket of foods consumed by US households, a GMO-free basket would increase the average family food budget by US$2719 annually (about a 28.7% increase from the US$9652 base).

Conventional model

Empirically, the conventional model is most characteristic of two separate categories of countries. One category comprises certain major agricultural countries such as Argentina, Brazil, Japan and New Zealand. This list is suggestive only, rather than exhaustive. These countries are concerned with both protection of domestic production traditions and non-GMO reputation for product export. The other category comprises agriculturally oriented developing countries in which hunger and poverty alleviation should be overriding concerns. These concerns overlap with domestic production traditions and non-GMO reputation for product export.

The conventional model of public opinion and public interest is shaped primarily by traditional consumer and producer values and also agricultural interests. This model is not necessarily precautionary in orientation. A consideration is that agricultural biotechnology is likely to be imported, as technology or as products – and agricultural biotechnology is dominated by US companies and scientists. The conventional model more closely resembles the situation of any industry facing import competition in its home market. Domestic consumption and production may be particularistic by country, and reflected in political influence with the government. The configuration of values, interests and perceived rights favors conventionalism (Walls et al., 2005). The public interest reflects a combination of local business interests and consumer preferences and values. There may be significant elements of tradition, as in wine production in France and Italy, where location and varietal are regarded as extremely important (as well as profitable). Human intervention in natural selection through breeding control

is different from gene splicing or cloning, or similar genetic modifications (GMOs).

Developing countries constitute a separable category, shaped by the problems of hunger and poverty alleviation, within the conventional model (Paarlberg, 2000; Gerasimova, 2016). There may be alternative solutions to a global food crisis (Fraser et al., 2016), such as reduction of food waste. Hunger concerns may or may not increase GMO receptivity (Carter et al., 2016). The hunger argument may collide with the preserving nature argument (Freedman, 2013), especially if there are alternative solutions for hunger alleviation. There are expressed reservations concerning whether GMOs in fact increase agricultural productivity yield (Hakim, 2016) and other alleged benefits (Greenpeace, 2016). An absence of increased productivity yield or nutritional benefits makes a moral argument favoring GMOs as hunger relief more awkward. Epstein (2014) argued a moral case for fossil fuels as elevating human welfare (Broughton, 2014): but the argument is that immediate abandonment of fossil fuels (without an equivalent economic cost alternative) would involve a substantial reduction in human welfare, especially in developing countries. It is not clear presently that a similar case can be made in favor of GMOs (Barrows et al., 2014).

Socio-economic effects of GMOs may be of greater importance to countries in the conventional model (Fischer et al., 2015; Binimelis and Myhr, 2016) and such effects are emphasized (Falck-Zepeda and Zambrano, 2011) in the Cartagena Protocol on Biosafety to the Convention on Biological Diversity. Governments and businesses assess social and economic consequences as more important than the GMO issue.

Anderson and Jackson (2004) analyzed potential economic effects of GMO crops in sub-Saharan Africa. The first generation aimed at increasing farmer profitability through some combination of cost reduction and increased yield. The second generation aimed additionally at consumer interests, such as Golden Rice. Anderson and Jackson concluded that there could be quite large welfare gains in Africa and that the welfare gains would be diminished only slightly by the EU's then anti-GMO import barriers. Country bans on GMO crop imports as a device for maintaining EU market access for non-GMO products would harm domestic consumers far more than the estimated small economic gain for African farmers through greater EU market access.

Accommodative model

The accommodative model means that public opinion is neither hostile to agricultural biotechnology nor inclined toward the European precautionary principle. US public opinion may be relatively positive toward science and technology. It has been argued that US press coverage tends to limit GMO controversy (Nisbet, 2006). Absent legislative action against agricultural biotechnology, two decisions are deferred to administrative/regulatory agencies: (i) should approval be granted for biotechnology procedures and agricultural/food products; (ii) how much information must be disclosed to consumers. Both decisions tend to be evaluated using agency cost–benefit analysis procedures and company/industry/public input procedures. While pro-precaution and pro-convention interests may argue against approval and pro-consumer interests may argue for maximum disclosure, the general context is absence of hostile public opinion and deferral to delegated administrative process. Public opinion tends to regard decisions as matters of elite/technical judgment that determines the public interest defined narrowly in terms of consumer safety. While environmental externality is recognized as a possibility (such as escape of GMO varieties into nature), judging that possibility is also deferred to delegated administrative process. An attribute of the accommodative model is that a vital scientific question is whether there is evidence indicating absence of safety: that is, there has not been disconfirmation of safety. The precautionary principle, discussed above, makes a different assumption that safety must be proved beyond something like reasonable doubt.

The accommodative model is not universal in the USA. In November 2016, Sonoma County, California – an important vineyard area – adopted a ballot measure making the county the largest GMO-free zone in the USA (Froelich, 2016). In November 2012, Proposition 37, a state-wide ballot measure requiring mandatory labeling, was narrowly rejected 51.4% to 48.6% (Bovay and Alston, 2016; Xu, 2016).

A key consideration is proper labeling of GMO products and GMO content (Center for Food Safety, 2016). Labeling requirements vary markedly across countries (Ghozzi et al., 2016). In the USA, mandatory labeling has been resisted by businesses (Galant, 2005). In July 2016, the US government enacted a federal GMO labeling statute, largely for the purpose of preempting variable state requirements (Lambrecht, 2016).

Wunderlich et al. (2016) surveyed 331 adult consumers and 17 supermarket representatives in northern New Jersey. The majority of the representatives believed that non-GMO labeling influenced consumer behavior. There was some distribution of opinion: 52.9% thought labeling affected all consumers; 17.6% restricted effect to knowledgeable and anti-GMO consumers; and 5.9% thought clientele not interested (as a result of specific demographics of that clientele). Of the adult consumer respondents, only 35.3% reported even noticing non-GMO labeling. The survey found only a moderate correlation between knowledge and purchasing behavior, but a stronger correlation between attitude and purchasing behavior.

Defining Country Context

The models are formulated as ideal-type descriptions representative of both the logical possibilities and descriptively specific real country contexts. Empirically, all three models might exist within a particular country, but one model is typically dominant. Moreover, the representative model for a particular country could shift due to change in dominant (or supermajority) public opinion and conception of the public interest. The three models can coexist both within countries and across regions of the world.

Each model can be depicted as a different configuration of six important dimensions that can vary across countries in determining which model is characteristic of the particular country. I use these six dimensions because they capture the essential features of the GMO controversy in ideal-type conceptualization. The six dimensions are domestic public policy, business, media, scientific experts, domestic public opinion and anti-GMO activists. How these six dimensions take definition and interact is part of each model. There are significant variations in how populations and businesses in specific countries understand agricultural biotechnology. Working out the public interest involves the specific interaction among the six dimensions varying by country.

The precautionary model is depicted in Fig. 5.1, the conventional model in Fig. 5.2 and the accommodative model in Fig. 5.3. Each figure is structured into three columns each of two dimensions, as viewed by the reader. In the center column, governments set public policies and businesses can undertake to design, make and/or use genetically engineered products and also lobby governments and influence public opinion. Businesses include agricultural enterprises that use genetically engineered products. In the left column, activists seek to influence the other dimensions, generally against GMOs;

and public opinion is strongly negative or neutral. In the right column, experts provide desirably neutral and validated scientific information. Media provide (or avoid providing) coverage of issues, debates and information. In each figure, the least influential dimensions are highlighted through italics. The influential dimensions are linked to public policy by arrows, while dimensions with limited influence are italicized.

Figure 5.1 depicts the precautionary model in the EU. Activists have strong influence, public opinion is negative toward GMOs and media is supportive of controversy or may be anti-GMO. Pro-GMO experts, shown in italics, have limited influence. European businesses have interests in opposing US GMOs. As a result, public policy is basically hostile to US GMOs. In the precautionary model, activists and pretty negative public opinion are dominant in influencing public policy. The ethical tension is whether human welfare effects are given sufficient attention, in a condition of business self-interest favoring conventional domestic business. The precautionary model is essentially global in outlook: if GMOs are undesirable in the EU, they are undesirable globally. This posture might arguably harm human welfare globally.

Figure 5.2 depicts the conventional model dominant in some agricultural countries.

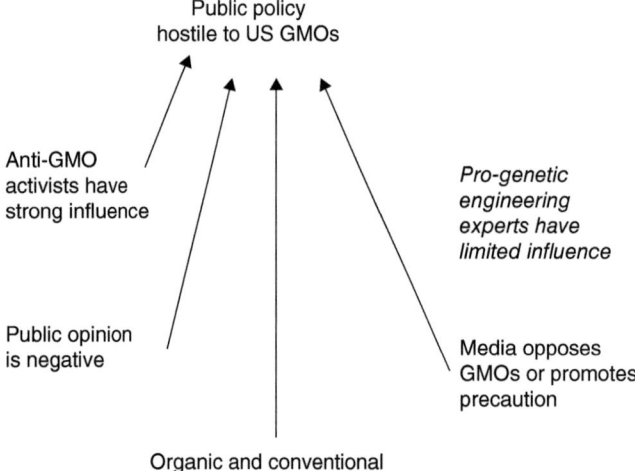

Fig. 5.1. The precautionary model in the European Union. (Duane Windsor.)

This model is pro-domestic business – which happens to be conventional – and pro-GMO experts are viewed as foreigners (and typically these foreigners are US based), shown in italics. Public opinion prefers domestic agriculture and anti-GMO activists have strong influence. Media supports domestic interests. Domestic businesses have favorable treatment in public policy. In the conventional model, non-genetic engineering domestic agricultural businesses are dominant in influencing public policy. The determinants and outcomes are much more country specific than in the precautionary or accommodative models. The ethical tension is very clearly between human welfare effects and domestic conventional agricultural interests. If domestic consumers would benefit from GMOs, then consumer welfare is arguably reduced by domestic business interests.

Figure 5.3 depicts the accommodative model in the USA. In this model, anti-GMO activists, shown in italics, have limited influence; and public opinion, also shown in italics for emphasis, is neutral. Pro-GMO experts are influential and media arguably suppresses controversy. Public policy thus tends to be favorable to GMOs. In the accommodative model, pro-genetic engineering experts and businesses are dominant in shaping public opinion and media coverage. Public opinion arguably remains reluctant concerning scientific confidence and media coverage arguably suppresses controversy. The ethical tension is that there may not be substantial human welfare gains even if there is reasonable

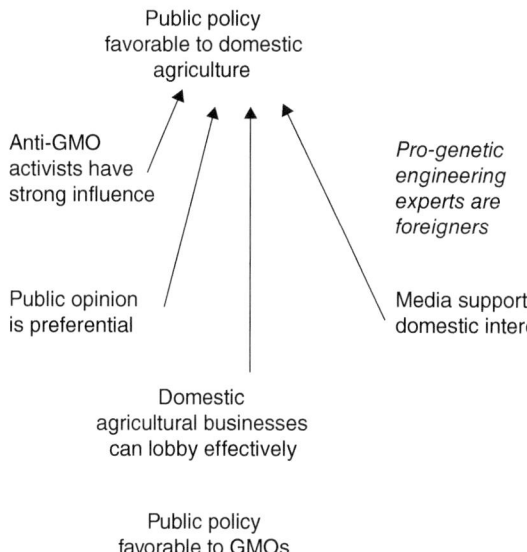

Fig. 5.2. The conventional model in some agricultural countries. (Duane Windsor.)

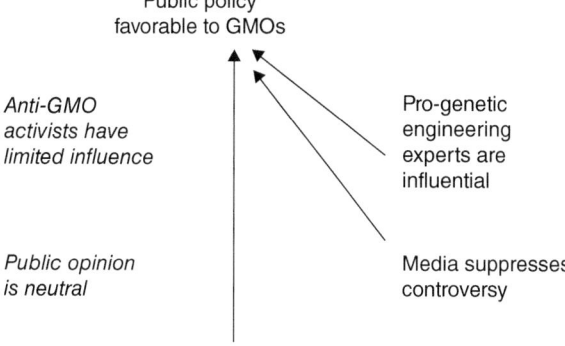

Fig. 5.3. The accommodative model in the USA. (Duane Windsor.)

GMO safety, and GMO agriculture is largely a substitute competing with conventional agriculture. Pro-GMO businesses are simply pursuing self-interest, while pro-GMO experts focus on safety as distinct from welfare effects more broadly defined.

Even the accommodative model in the USA is not strictly speaking an endorsement model. Rather, public opinion is neutral and media arguably suppresses controversy. A secondary contribution of the chapter is to explain why a full-blown GMO endorsement model – which the accommodative model is not – does not exist presently. An endorsement model can be defined logically as an ideal-type, but not associated empirically with a specific country context. The endorsement model is simply Fig. 5.3 with no italicized dimensions and thus anti-GMO activism essentially disappears and public opinion becomes supportive, with an additional arrow from public opinion to public policy. The enthusiastic acceptance and endorsement of agricultural and medical biotechnology by virtually everyone simply does not exist. The US accommodative model involves essentially neutral public opinion deferring to scientific experts and delegating public policy to government, in a condition in which media arguably suppresses controversy. What the three proposed models help explain empirically is why the endorsement model does not exist. The endorsement model is simple to state: all key stakeholders are pro-GMO, the benefits of which are readily apparent to most and scientifically validated. Increased competition among businesses and scientists in developing and producing GMO products is not strictly necessary, but seems more likely than not to occur over time together with pro-GMO learning by stakeholders. Longer-term movement from precaution toward accommodation may be quite gradual and will involve learning about risks, benefits and scientific evidence. A survey on Facebook and Twitter for a small respondent count (n = 150) found that most individuals are open to learning more about GMOs, while believing that GMOs should be heavily monitored (Paz and Zhang, 2018). Movement, at whatever speed, should be assessed in context: robotics, artificially intelligent systems and alternative energy sources, including possibly more nuclear power supply, will also likely be increasing in the future.

Conclusion

The primary purpose of this chapter is to address the normative research question of whether pro-GMO business and science should lead or follow public opinion. The ethical tension involves business and government choices in light of the gap among public opinion, business interest and expert advice. The evidence concerning GMO safety can be disputed and the human welfare improvements from GMO agricultural products are not yet well established. In this gap, GMO experts emphasize safety but pro-GMO businesses may be engaged simply in self-interest. The EU precautionary model and the conventional agriculture model found in some countries are quite different from the accommodative model in the USA. Even the accommodative model is not strictly speaking an endorsement model.

The future will be increasingly an engineered world, including genetically engineered foods and pharmaceuticals (Jahren, 2016), as well as animals and perhaps humans. An endorsement model may be coming, but since this longer run is difficult to predict in any detail, the proposed set of accommodative, conventional and precautionary models may continue to have explanatory merit. Robotics and artificial intelligence will also be greatly developed. Scientific knowledge and engineering innovation tend to prove irresistible. While proper balance among risk, precaution and science (Stirling, 2007) remains unresolved, the general forces of technological innovation, desirable alleviation of hunger, poverty and disease, trade globalization, and favorable scientific consensus may tend to reduce perception of risk and work against precaution. If so, then in Fig. 5.1, the future is a movement away from precaution and toward accommodation. An educated guess would be that engineered food consumption could tend to lag behind other engineered sectors such as pharmaceuticals, animals and humans. Some consumers and some agricultural producers will prefer conventional or organic food products, for a variety of reasons. Success in other engineered sectors will tend to soften resistance and promote acceptance as generations change over time. The Greenpeace posture against Golden Rice, noted earlier, has its merits in stressing precaution, corporate lobbying, absence of strong evidence of desirability and local option to choose conventional agriculture. Nevertheless, the Greenpeace posture may rest on a presumption of a pristine natural and human environment that in a sense no longer exists, except in arguably defensible local pockets (including both the EU and various developing countries).

A wide gap between public opinion and scientific consensus is not due solely, or even largely, to anti-GMO activism (Fluegge, 2016) and EU precaution. As Fluegge points out, the requirement for influencing public opinion is not reliance on expert scientific consensus, because there is arguably strong public resistance to simple assertion of safety (by experts and businesses). Fluegge argues that persuasion must occur at the level of the individual consumer, in some process analogous to patient-centered medicine. It is the individual who bears the risk of consumption if the experts are wrong and businesses are driven by profit motive rather than customer safety. The individual consumer has reservations about whether to bear risk, potentially long term in occurrence (like cancer from tobacco consumption), on expert advice that might be less than 100% and strictly certain. If so, then it is the individual hesitation, reinforced by activism and precaution, that underlies the precautionary model and resists the accommodative model. The conventional model, in contrast, is more shaped by country-specific interests, including agricultural business interests in a globalizing trade-oriented world economy.

The 'responsible governance' design problem for genetic engineering (Hartley *et al.*, 2016) involves the ethics and rhetoric of debate (Devos *et al.*, 2008; Preston and Wickson, 2016) and learning over time (Hansson, 2016). The three models are aligned in a way that suggests movement over time from precaution toward accommodation. Hansson suggests that, in the language of this chapter, precaution and accommodation may need to be tailored to changing scientific evidence concerning risk. If evidence begins to identify greater potential danger than the scientific consensus has suggested, precautionary measures, science based, will need to be strengthened. If evidence begins to reinforce the scientific consensus, precautionary measures, science based, will need to be reduced. The capacity to tailor in this way depends on public opinion and the goals and strategies of anti-GMO activists. The conventional model – varying by country – is more dependent on country-specific business considerations.

References

American Association for the Advancement of Science (2012) Statement by the AAAS Board of Directors on labeling of genetically modified foods. Available at: http://www.aaas.org/sites/default/files/AAAS_GM_statement.pdf (accessed 21 November 2017).

Anderson, K. and Jackson, L.A. (2004) Implications of genetically modified food technology policies for sub-Saharan Africa. World Bank Policy Research Working Paper No. 3411. Available at: http://papers.ssrn.com/paper.taf?abstract_id=625288 (accessed 21 November 2017).

Barrows, G., Sexton, S. and Zilberman, D. (2014) Agricultural biotechnology: The promise and prospects of genetically modified crops. *Journal of Economic Perspectives* 28, 99–120. doi:10.1257/jep.28.1.99

Bartz, D. and Roumeliotis, G. (2016) Bayer's Monsanto acquisition to face politically charged scrutiny. *Reuters*, 15 September. Available at: https://www.reuters.com/article/us-monsanto-m-a-bayer-antitrust/bayers-monsanto-acquisition-to-face-politically-charged-scrutiny-idUSKCN11K2LG (accessed 21 November 2017).

Baumol, W.J. (1976/2016) Smith vs. Marx on business morality and the social interest. *The American Economist* 61, 44–51. doi:10.1177/0569434515627110

Binimelis, R. and Myhr, A.I. (2016) Inclusion and implementation of socio-economic considerations in GMO regulations: Needs and recommendations. *Sustainability* 8, Article 62. doi:10.3390/su8010062

Bodiguel, L. (2016) GMO, conventional and organic crops: From coexistence to local governance. *Agriculture and Agricultural Science Procedia* 8, 263-269. doi:10.1016/j.aaspro.2016.02.102

Bovay, J. and Alston, J.M. (2016) GM labeling regulation by plebiscite: Analysis of voting on Proposition 37 in California. *Journal of Agricultural and Resource Economics* 41, 161–188. Available at: http://www.waeaonline.org/UserFiles/file/JAREMay20161Bovay161-188.pdf (accessed 4 July 2018).

Bratspies, R. (2004) Consuming (f)ears of corn: Public health and biopharming. *American Journal of Law and Medicine* 30, 371–404. doi:10.1177/009885880403000211

Broude, T. (2006) Genetically modified rules: The awkward rule-exception-right distinction in EC-biotech. Hebrew University International Law Research Paper No. 14-06. Available at: https://papers.ssrn.com/sol3/papers.cfm?abstract_id=949623 (accessed 1 November 2017).

Broughton, P.D. (2014) Making 'the moral case for fossil fuels': Renouncing oil and its byproducts would plunge civilization into a pre-industrial hell – a fact developing countries keenly realize. Available at: http://www.wsj.com/articles/book-review-the-moral-case-for-fossil-fuels-by-alex-epstein-1417477909 (accessed 21 November 2017).

Bukenya, J.O. and Wright, N.R. (2007) Determinants of consumer attitudes and purchase intentions with regard to genetically modified tomatoes. *Agribusiness* 23, 117–130. doi:10.1002/agr.20109

Carbonell, I.M. (2016) The ethics of big data in big agriculture. *Internet Policy Review* 5. doi:10.14763/2016.1.405

Carter, B.E., Conn, C.C. and Wiles, J.R. (2016) Concern about hunger may increase receptivity to GMOs. *Trends in Plant Science* 21, 539–541. doi:10.1016/j.tplants.2016.05.003

Center for Food Safety (2016) Issues: GE Food Labeling. Available at: http://www.centerforfoodsafety.org/issues/976/ge-food-labeling (accessed 28 November 2016).

Convention on Biological Diversity (1992) Article 2. Use of terms. Available at: https://www.cbd.int/convention/articles/default.shtml?a=cbd-02 (accessed 20 November 2017).

Corlett, R.T. (2017) A bigger toolbox: Biotechnology in biodiversity conservation. *Trends in Biotechnology* 35, 55–65. doi:10.1016/j.tibtech.2016.06.009

Devos, Y. *et al.* (2008) Ethics in the societal debate on genetically modified organisms: A (re) quest for sense and sensibility. *Journal of Agricultural and Environmental Ethics* 21, 29–61. doi:10.1007/s10806-007-9057-6

Disdier, A.-C. and Fontagne, L. (2009) Trade impact of European measures on GMOs condemned by the WTO panel. CEPII Working Paper No. 2009-16. Available at: https://papers.ssrn.com/sol3/papers.cfm?abstract_id=2004984 (accessed 1 November 2017).

DowDuPont (2016) DowDuPont Merger of Equals Update. Filing with SEC, 16 June. Available at: https://perma.cc/X2ST-FUBL (accessed 21 November 2017).

Epstein, A. (2014) *The Moral Case for Fossil Fuels*. Portfolio/Penguin, New York.

European Commission (2015) Reviewing the decision-making process on genetically modified organisms (GMOs). Available at: http://eur-lex.europa.eu/resource.html?uri=cellar:3cc2bfe5-e9d1-11e4-892c-01aa75ed71a1.0002.03/DOC_1&format=PDF (accessed 21 November 2017).

European Food Safety Authority (2013) Guidance on the environmental risk assessment of genetically modified animals. *EFSA Journal* 11, 3200, 1–190. doi:10.2903/j.efsa.2013.3200

Fahlgren, N., Bart, R., Herrera-Estrella, L., Rellán-Álvarez, R., Chitwood, D.H. *et al.* (2016) Plant scientists: GM technology is safe. *Science* 351, 824. doi:10.1126/science.351.6275.824-a

Fairfield-Sonn, J.W. (2016) Political economy of GMO foods. *Journal of Management Policy and Practice* 17, 60–70. Available at: http://www.na-businesspress.com/JMPP/Fairfield-SonnJW_Web17_1_.pdf (accessed 4 July 2018).

Falck-Zepeda, J.B. and Zambrano, P. (2011) Socio-economic considerations in biosafety and biotechnology decision making: The Cartagena Protocol and national biosafety frameworks. *Review of Policy Research* 28, 171–195. doi:10.1111/j.1541-1338.2011.00488.x

Ferer, B.S. (2016) The European Commission's GMO opt-out for member states: A WTO perspective. *European Journal of Risk Regulation* 7, 187–190. doi:10.1017/S1867299X00005481

Fischer, K., Ekener-Petersen, E., Rydhmer, L. and Björnberg, K.E. (2015) Social impacts of GM crops in agriculture: A systematic literature review. *Sustainability* 7, 8598–8620. doi:10.3390/su7078598

Fluegge, K. (2016) Social learning theory and public perception of GMOs: What Blancke et al. (2015) and other plant biotechnologists are missing. *Journal of the Science of Food and Agriculture* 96, 2939–2940. doi:10.1002/jsfa.7634

Fraser, E. *et al.* (2016) Biotechnology or organic? Extensive or intensive? Global or local? A critical review of potential pathways to resolve the global food crisis. *Trends in Food Science & Technology* 48, 78–87. doi:10.1016/j.tifs.2015.11.006

Freedman, D.H. (2013) The truth about genetically modified food. *Scientific American*, 1 September. Available at: https://www.scientificamerican.com/article/the-truth-about-genetically-modified-food/ (accessed 21 November 2017).

Froelich, A. (2016) Sonoma County [California] voter success leads to largest GMO-free zone in U.S. Available at: http://www.trueactivist.com/sonoma-county-voter-success-leads-to-largest-gmo-free-zone-in-u-s/ (accessed 21 November 2017).

Funk, C. and Rainie, L. (2015) Americans, politics and science issues. Available at: http://www.pewinternet.org/2015/07/01/americans-politics-and-science-issues/ (accessed 21 November 2017).

Galant, C.R. (2005) Labeling limbo: Why genetically modified foods continue to duck mandatory disclosure. *Houston Law Review* 42, 125–164. Available at: http://www.houstonlawreview.org/archive/downloads/42-1_pdf/Galant.pdf (accessed 4 July 2018).

Gerasimova, K. (2016) Debates on genetically modified crops in the context of sustainable development. *Science and Engineering Ethics* 22, 525–547. doi:10.1007/s11948-015-9656-y

Ghozzi, H., Soregaroli, C., Boccaletti, S. and Sauvée, L. (2016) Impacts of non-GMO standards on poultry supply chain governance: Transaction cost approach vs. resource based view. *Supply Chain Management: An International Journal* 21, 743–758. doi:10.1108/SCM-03-2016-0089

Goodwin, B.K., Marra, M.C. and Piggott, N.E. (2016) The cost of a GMO-free market basket of food in the United States. *AgBioForum* 19, Article 3. Available at: http://www.agbioforum.org/v19n1/v19n1a03-marra.htm (accessed 4 July 2018).

Greenpeace (2016) Golden Rice. Available at: http://www.greenpeace.org/international/en/campaigns/agriculture/problem/Greenpeace-and-Golden-Rice/ (accessed 8 December 2016).

Hakim, D. (2016) Debate over genetic crops' worth rages: Study shows agriculture modification hasn't produced the promised bounty. *Houston Chronicle*, 30 October, 116, A34.

Hansson, S.O. (2016) How to be cautious but open to learning: Time to update biotechnology and GMO legislation. *Risk Analysis* 36, 1513–1517. doi:10.1111/risa.12647

Hartley, S., Gillund, F., van Hove, L. and Wickson, F. (2016) Essential features of responsible governance of agricultural biotechnology. *PLoS Biol* 14, e1002453. doi:10.1371/journal.pbio.1002453

Heald, P.J. and Smith, J.C. (2006) The problem of social cost in a genetically modified age. *Hastings Law Journal* 58, 87–151. Available at: http://digitalcommons.law.uga.edu/cgi/viewcontent. cgi?article=1007&context=fac_artchop (accessed 4 July 2018).

Herring, R. and Paarlberg, R. (2016) The political economy of biotechnology. *Annual Review of Resource Economics* 8, 397–416. doi:0.1146/annurev-resource-100815-095506

Jahren, H. (2016) Engineered food holds our future. *Time*, 24 October, 188, 78–79.

Kangmennaang, J., Osei, L., Armah, F.A. and Luginaah, I. (2016) Genetically modified organisms and the age of (un) reason? A critical examination of the rhetoric in the GMO public policy debates in Ghana. *Futures* 83, 37–49. doi:10.1016/j.futures.2016.03.002

Kelman, S. (1981) Cost-benefit analysis: An ethical critique. *AEI Journal on Government and Society Regulation* Jan/Feb, 33–40. Available at: https://object.cato.org/sites/cato.org/files/serials/files/regulation/1981/1/v5n1-7.pdf (accessed 4 July 2018).

Lambrecht, B. (2016) Congress puts rush on GMO labeling: If passed, federal law will overrule each state's rules. *Houston Chronicle*, 11 July, 115, A1, A14.

Levidow, L. and Carr, S. (2000) Unsound science? Transatlantic regulatory disputes over GM crops. *International Journal of Biotechnology* 2, 257–273. doi:10.1504/IJBT.2000.000131

Livermore, M.A. (2014) Cost-benefit analysis and agency independence. *University of Chicago Law Review* 81, 609–688. Available at: http://www.thecre.com/pdf/20170827_livermore.pdf (accessed 4 July 2018).

Lyon, J. (2016) Nobel laureates pick food fight with GMO foes. *JAMA* 316, 1752–1753. doi:10.1001/jama.2016.11571

Malyska, A., Bolla, R. and Twardowski, T. (2016) The role of public opinion in shaping trajectories of agricultural biotechnology. *Trends in Biotechnology* 34, 530–534. doi:10.1016/j.tibtech.2016.03.005

Matos, C.A. de (2004) Consumer attitudes toward genetically modified foods: An experimental approach. Available at: https://papers.ssrn.com/sol3/papers.cfm?abstract_id=599302 (accessed 1 November 2017).

National Academies of Sciences, Engineering, and Medicine (2016) *Genetically Engineered Crops: Experiences and Prospects*. NAS, Washington, DC. doi:10.17226/23395

Nicolia, A., Manzo, A., Veronesi, F. and Rosellini, D. (2014) An overview of the last 10 years of genetically engineered crop safety research. *Critical Review of Biotechnology* 34, 77–88. doi:10.3109/07388551.2013.823595

Nisbet, M. (2006) How press coverage limits controversy in the U.S. over plant biotechnology. The Committee for Skeptical Inquiry, 22 February.

Available at: http://www.csicop.org/specialarticles/show/how_press_coverage_limits_controversy_in_the_u.s._over_plant_biotechnology (accessed 21 November 2017).

Paarlberg, R. (2000) Agrobiotechnology choices in developing countries. *International Journal of Biotechnology* 2, 164–173. doi:10.1504/IJBT.2000.000134

Panchin, A.Y. and Tuzhikov, A.I. (2017) Published GMO studies find no evidence of harm when corrected for multiple comparisons. *Critical Reviews in Biotechnology* 37, 213–217. doi:10.3109/07388551.2015.1130684

Paz, A.M. and Zhang, X.-N. (2018) The GMO industry: A neglected earthly frontier. *Journal of Hunger & Environmental Nutrition* 13, 277–288. doi:10.1080/19320248.2016.1227755

Perez, O. (2006) Anomalies at the Precautionary Kingdom: Reflections on the GMO Panel's decision. Hebrew University International Law Research Paper No. 13-06. Available at: https://papers.ssrn.com/sol3/papers.cfm?abstract_id=940907 (accessed 1 November 2017).

Preston, C.J. and Wickson, F. (2016) Broadening the lens for the governance of emerging technologies: Care ethics and agricultural biotechnology. *Technology in Society* 45, 48–57. doi:10.1016/j.techsoc.2016.03.001

Ramsey, L. and Brodwin, E. (2015) The real reason Americans are worried about GMOs is actually a way bigger problem. Available at: http://www.businessinsider.com/americans-doubt-scientists-on-gmos-2015-7 (accessed 21 November 2017).

Renn, O. (2007) Precaution and analysis: two sides of the same coin? *EMBO Reports* 8, 303–304. doi:10.1038/sj.embor.7400950

Reuters (2016a) EU resumes investigation into DuPont Dow merger deal. *Reuters*, 3 October. Available at: http://www.reuters.com/article/us-dupont-m-a-dowchemical-eu-idUSKCN1230VE (accessed 21 November 2017).

Reuters (2016b) Bayer clinches Monsanto with improved $66 billion bid. *Reuters*, 14 September. Available at: http://www.cnbc.com/2016/09/14/bayer-and-monsanto-agree-to-merge.html (accessed 21 November 2017).

Robaey, Z. (2016) Gone with the wind: Conceiving of moral responsibility in the case of GMO

contamination. *Science and Engineering Ethics* 22, 889–906. doi:10.1007/s11948-015-9744-z

Sindico, F. (2005) The GMO dispute before the WTO: Legal implications for the trade and environment debate. FEEM Working Paper No. 11.05. Available at: https://papers.ssrn.com/sol3/papers.cfm?abstract_id=655061 (accessed 1 November 2017).

Singh, M., Bhoge, R.K. and Randhawa, G. (2016) Crop-specific GMO matrix-multiplex PCR: A cost-efficient screening strategy for genetically modified maize and cotton events approved globally. *Food Control* 70, 271–280. doi:10.1016/j.foodcont.2016.05.032

Stirling, A. (2007) Risk, precaution and science: Towards a more constructive policy debate. Talking point on the precautionary principle. *EMBO Reports* 8, 309–315. doi:10.1038/sj.embor.7400953

Tagliabue, G. (2016a) The necessary 'GMO' denialism and scientific consensus. *Journal of Science Communication* 15, Y01-1. https://jcom.sissa.it/sites/default/files/documents/JCOM_1504_2016_Y01.pdf

Tagliabue, G. (2016b) European incoherence on GMO cultivation versus importation. *Nature Biotechnology* 34, 694–695. doi:10.1038/nbt.3588

Wallace, D.L. (2016) A potential liability scenario for 'non-GMO' labeling. *HSF Legal Briefings*. Available at: http://works.bepress.com/david_wallace1/19/ (accessed 1 November 2017).

Walls, J., Rogers-Hayden, T., Mohr, A. and O'Riordan, T. (2005) Seeking citizens' views on GM crops: Experiences from the United Kingdom, Australia, and New Zealand. *Environment: Science and Policy for Sustainable Development* 47, 22–37. doi:10.3200/ENVT.47.7.22-37

Weih, M. and Polle, A. (2016) Editorial: Ecological consequences of biodiversity and biotechnology in agriculture and forestry. *Frontiers in Plant Science* 7, 210. doi:10.3389/fpls.2016.00210

Wunderlich, S.M., Gatto, K.A. and Vecchione, M. (2016) Labeling of genetically modified food products and consumer behavior. *The FASEB Journal* 30, 1151.11. Available at: https://www.fasebj.org/doi/10.1096/fasebj.30.1_Supplement.1151.11 (accessed 4 July 2018).

Xu, S.M. (2016) Who wants the right to know? An analysis of GMO-labeling in California. *Journal of Environmental and Resource Economics at Colby* 3, Article 5. Available at: http://digitalcommons.colby.edu/jerec/vol3/iss1/5 (accessed 4 July 2018).

6 Genetically Modified Organisms in Food: Ethical Tensions and the Labeling Initiative

Debra M. Strauss*

Charles F. Dolan School of Business, Fairfield University, Fairfield, Connecticut, USA

Introduction

This chapter explores the ethical implications of genetically modified organisms (GMOs) in food that originate from plants genetically altered through bioengineering. The tensions between the proliferation of agricultural biotechnology and consumer concerns about potential harm to human health and the environment ultimately cause us to reflect on the current regulatory scheme in the USA. Does the failure to require adequate, meaningful labeling, and the preemption of grassroots efforts to do so, violate our right to informed consent by not allowing consumers a choice as to whether to knowingly and willingly assume the risks of ingesting GMOs?

Genetically modified (GM) plants involve a uniquely invasive application of agricultural biotechnology, unlike traditional plant breeding and hybrid methods used in the past. Through this novel process, the DNA of one organism is inserted into another, causing the target trait to be expressed in that non-related species at the cellular level throughout the plant, including the fruit or vegetable and the component ingredients that become part of a variety of food products. Most commonly, GM plants are engineered to withstand a weed-killing pesticide, Roundup, sold by Monsanto along with the herbicide-resistant varieties of soybeans, canola, cotton, corn, radicchio, rice and sugar beet. In addition, genes derived from a bacterium in the soil used as an insecticide, *Bacillus thuringiensis* (Bt), have been inserted into crops to induce the plant to produce a toxin against certain insects, producing Bt-corn, Bt-cotton, Bt-potatoes, Bt-rice and Bt-tomatoes (Strauss, 2006).

There is even a biotechnology invention, as yet undeveloped for commercial use due to widespread consumer opposition in the international community, called the 'Terminator' gene, also known as Genetic Use Restriction Technologies (GURTs). Developed jointly by the agricultural biotechnology industry and the United States Department of Agriculture (USDA), this technology genetically alters plants to produce sterile seeds at harvest, effectively blocking the ability of the plant to procreate future generations in order to prevent farmers' traditional practice of saving and replanting harvested seed and thereby necessitating the purchase of new seeds from the biotechnology company each year (Strauss, 2009). Recognizing the conceivable catastrophic effect on the global food supply if spread via common cross-pollination, the United Nations, through the UN Convention on Biological Diversity in 2000, implemented and has continued to maintain a de facto moratorium on these sterile seed technologies (Convention on Biological Diversity, 2017). Ultimately,

* E-mail: DStrauss@fairfield.edu

the very existence of the Terminator serves as a reminder that there is virtually no limit as to what might be developed in the future through this novel technology.

In the USA, GM crops now comprise almost all of the plantings and the vast majority of component products in the US market (Strauss, 2012a). Recent statistics indicate that GM crops accounted for 94% of soybeans, 92% of corn and 96% of cotton planted in 2017; and GM canola, squash, papaya, alfalfa and sugar beet were widely planted (USDA, 2017a). The Grocery Manufacturers Association (GMA) reported in 2005 that 75% of all processed foods in the USA contained a GM ingredient, including almost every product with a corn or soy ingredient and some with canola or cottonseed oil; this figure has continued to be widely cited. However, projecting from the exponentially higher plantings since then, the prevalence should be substantially higher today. The current GMA position statement emphasizes the proliferation of GMOs:

> It is important for our consumers to know that this technology is not new. In fact, it has been around for the past 20 years, and today, 70–80% of the foods we eat in the United States, both at home and away from home, contain ingredients that have been genetically modified. If the ingredient label on any food or beverage product contains corn or soy, they most likely contain genetically modified ingredients, as a very high percentage of those crops grown in the U.S. use GM technology. In addition, a high percentage of other ingredients in the U.S., such as sugar beets, are grown with the use of GM technology as well. (2017, n.p.)

Yet the widespread use of GMOs is unknown to most consumers because, unlike the European Union and the broader international community where more than 60 countries require labeling, the USA does not dictate mandatory labeling, rigorous approval or monitoring of GM plants and foods (Strauss, 2006; Consumer Reports, 2014). Moreover, due to the absence of long-term studies, the level of safety and the effect on human health and the environment remain uncertain.

Surveys on GM foods reveal that a significant majority of consumers believe that the government should include ethical and moral considerations when making regulatory decisions about genetic engineering. Moreover, consumers seek an active role from regulators to ensure that new products are safe (Strauss, 2007). Consumer polls uniformly demonstrate that a vast majority of US citizens would like GMOs to be labeled (Center for Food Safety, 2017). For instance, a recent Consumer Reports National Research Center survey found that 92% of Americans prefer GM foods to be labeled and more than 70% indicated that they do not want GMOs in their food (Consumer Reports, 2014). This is a critical area where examining the ethical implications can lead to further developments in the law as a means for the community to address and resolve these issues.

The juxtaposition between the proliferation of agricultural biotechnology and consumer concerns will be explored in view of the critical role of food safety in human health and the environment. To this end, the next section examines the key ethical issues arising from genetic engineering, particularly the right to informed consent and conflicts of interest in studies and scientific research that may hamper inquiry into the possible long-term risks. The chapter then analyzes the new federal statute recently passed in the USA purportedly to label these foods and evaluates whether in fact it does so in an effective and meaningful way to fulfill these consumers' rights. The chapter highlights an approach that takes into account these ethical tensions and shifts the dialogue from public outcry to the policy and regulation arena. Accordingly, this proposal embraces participation of all stakeholders, education of the public on the potential risks, development of comprehensive labeling to enable informed consumer choice and establishment of a more active and independent role for government agencies in regulating biotechnology companies. The chapter concludes by providing queries to assist policy makers in implementing a new regulatory framework in the USA, recognizing that consideration be given to ethical tensions in shaping the international policy arena.

Ethical Tensions in Connection with GMOs

To recognize an ethical perspective in the ongoing discussions about GMOs, the Food and Agriculture Organization of the United Nations (FAO) formed an expert panel on ethics in food and agriculture, which met in four sessions and issued reports on specific focus areas (FAO, 2007a). The first report of the FAO panel, 'Ethical issues in food and agriculture', introduced ethical questions related to its mandate, such as: What is the value of food? What is the value of human health? What is the value of nature and natural resources? The FAO panel recognized as principles the right to adequate food, optimization, trust, equity and informed consent in identifying these ethical concerns as central to the debate about the future (FAO, 2001a).

The FAO's second report, 'Genetically modified organisms, consumers, food safety and the environment', stressed the role of ethical considerations in food and agriculture with regard to GMOs as well as food safety and the environment (FAO, 2001b). Issues examined included ownership of the necessary tools to produce GMOs, potential consequences of their use and undesirable effects that could result from their application, both now and in the future. Most important, the report advocated the participation of all stakeholders in making decisions regarding GMOs, emphasizing that '[w]idely communicated, accurate and objective assessments of the benefits and risks associated with the use of genetic technologies should involve all stakeholders ... Experts have the ethical obligation to be proactive and to communicate in terms that can be understood by the lay person' (2001b, p. 25).

The FAO's third report delved further into the risks of GMOs and the ethical imperative to make human health and the environment the paramount concern and accordingly restrict the use of this technology:

> Genetic engineering introduces a new uncertainty, as it affects the genetic design of plants and animals and thus the composition of our food. The Panel reiterated the concern expressed in its earlier sessions over the potential environmental, health and socio-economic impacts of genetic engineering and genetically modified organisms (GMOs). Unless there is an overriding advantage that is apparent, the Precautionary Principle, which is the foundation of the Cartagena Protocol on Biosafety, would point to a preference for non-genetically engineered food. Even when there is an obvious advantage, an exhaustive testing of the safety of the genetically engineered food is required. (2005, p. 4)

In its fourth and final report, the FAO continued to emphasize

> the ethical requirement to avoid the risks of, while sharing the benefits of, biotechnologies as part of the advancement of science, which also involves an examination of the ethical issues related to intellectual property rights ... Ethical considerations of decision-making in relation to genetically modified organisms from the perspective of the consumer, food safety and the environment are closely related to this issue. (2007b, p. 10)

Thus, a discussion of ethical tensions is warranted in the public arena, and all of the stakeholders should have a voice in determining the policy decisions that consequently ensue.

These considerations have induced the European Union and other countries in accordance with the precautionary principle to limit the use of bioengineered foods and require labeling of foods with GM ingredients. The continued development of genetically modified plants raises several moral principles, such as respect for nature and the value of life; consideration of the environment; and equity, power and the economically disadvantaged (Strauss, 2007). The most significant ethical tensions will be explored further below: the right to informed consent and the latent impact of conflicts of interest in scientific research.

The right to informed consent

The foregoing reluctance of the US government to establish a rigorous labeling scheme for the treatment of GM crops and GMOs in food raises critical ethical tensions. For consumers, the lack of clear and meaningful disclosure of the fact that their food was developed using bioengineering techniques violates the right of informed consent. From an ethical perspective, particularly a Kantian model, US citizens have been deprived of their autonomy and freedom of choice (Strauss, 2007). Individuals have the fundamental right to know what they are buying and eating before making a purchasing decision.

An economic model also supports transparency and disclosure of this information (Brussel, 2003). According to this reasoning, 'the market for GMOs at both the consumer and producer level is unable to achieve a rational, efficient and socially optimal result due to asymmetrical information' (p. 430). Without adequate information, consumers cannot make rational decisions about whether to purchase and consume GMOs, farmers do not have the tools to negotiate with biotech seed producers and organic farmers cannot effectively allocate resources to protect their crops from contamination by genetic drift. This market can only function efficiently

> if a mechanism is established for ensuring that rational, scientifically-based information on the effects of GMOs on human health, agricultural production, and the environment is available to the public. Because transaction costs would be prohibitively high for individual consumers or farmers to obtain such information, a system of mandatory disclosures tied to discretionary participation in the market for GMOs should be established by the government. (Brussel, p. 432)

The government has a responsibility to protect its citizens, particularly in such a critical area as the safety of the food supply. As a matter of ethics, the as yet unknown risks must not be placed on the unsuspecting public rather than on the companies who have created these genetic modifications. To do so would betray consumers' trust in their government to ensure their health and well-being as fiduciaries acting on their behalf. The Food and Drug Administration (FDA) has recognized this mandate in its regulatory approach to other areas of the food supply, for example,

applying a zero-risk policy to prohibit the introduction into the food supply of food and color additives determined to cause cancer in laboratory animals (Strauss, 1987).

Moreover, in an area where the FDA perceives there to be no safety risk – food treated with ionizing radiation – it has nonetheless required mandatory labeling of such foods, with the international (Radura) symbol for radiation along with the statement 'Treated with radiation' or 'Treated by irradiation' on the food label (FDA, 2016). In mandating a disclosure on all irradiated foods, the FDA was cognizant of widespread consumer concerns about food irradiation. According to the agency, 'the large number of consumer comments requesting retail labeling attest to the significance placed on such information by consumers' (Strauss, 2006, p. 184; United States, 2012). Yet the FDA has not applied this reasoning for GMOs in food, where only nonbinding recommendations for voluntary labeling have been its policy (FDA, 2015).

Consumers have the right to choose what they eat, and informed choice can only be realized through mandatory labeling on the package that is accessible and understandable. According to Consumers International, consumers' desires and opinions should be respected due to a fundamental right to know and make informed decisions (Halloran and Hansen, 1999). For example, a lack of labeling as to the presence of an introduced gene removes the individual's right to avoid known allergens and control their own fate. Eight percent of children in the USA have food allergies, some of which can be fatal (Kolehmainen, 2001). When Pioneer Hi-Bred spliced Brazil nut genes into a soybean to improve its protein content, the altered soybean provoked severe allergic attacks in eight individuals sensitive to Brazil nuts but not soybeans (Nordlee et al., 1996). Without a label alerting consumers that a soybean could contain genes from a highly allergic nut, even individuals aware of their severe allergies would have no warning. As a matter of policy, vital information about the transgenic processing must be made available to those individuals who could be affected by important health risks (Nestle, 1996).

While the potential risks generate a need for labeling of the presence of GMOs, such an approach is also required beyond safety issues, as a matter of taste and preference and for many health-related reasons. It must be recognized that many consumers make food choices based on religious, ethical and environmental considerations, for example, deciding not to eat veal, mass-produced chickens or non-organic produce. If biotechnology raises similar ethical, health and environmental concerns, it is not irrational for people to act on these preferences and aversions to risk (Teitel and Wilson, 1999). In order to make these informed decisions, food products must be effectively labeled. As a matter of ethics and public policy, '[s]ince labeling laws are created to meet consumer needs, consumer opinion should be respected' (Halloran and Hansen, 1999, n.p.).

The decision to allow the public to consume unlabeled genetically engineered (GE) food strikes some people as 'grossly undemocratic and slanted too far in favor of corporate interests'. 'Should our society allow the purported commercial rights of a corporation to supersede the citizen's right to make informed decisions in the marketplace?' (Teitel and Wilson, p. 61). 'Every person has a right to make choices about what they eat. Every person has a right to know' (Strauss, 2007, p. 28). With an increasing crescendo of proponents, this concept has been building momentum as a 'Consumer Right to Know' policy grounded on a number of concerns apart from health and safety, including religious, ethical, dietary restrictions and environmental objections (Keane, 2006; Nauheim, 2009; Begley, 2017).

A recent study by the National Academies of Sciences (NAS, 2016) provided support for labeling of GMOs, not based on safety concerns but expressly on the ethical grounds of the consumers' right to know policies that respect consumer autonomy and fairness, reasoning that: 'if non-GE labeling is voluntary, many products would have no label information about GE content. Consumers would not know whether the product contained GE ingredients and so would be deprived of the ability to make an informed choice about each product' (2016, pp. 305–306). In its most significant statement, the NAS committee concluded:

Mandatory labeling provides the opportunity for consumers to make their own personal risk-benefit decisions (regardless of the regulatory determination of safety) and to express a preference for a method of production. A voluntary non-GE label places the burden on consumers who want to avoid GE foods to search for non-GE products and provides no information to consumers who may not be actively searching for the information but who might be informed by the label. Voluntary labeling also may not help consumers who cannot afford the kinds of foods that will be voluntarily labeled. (2016, p. 306)

For consumers who are careful about the content of the food they eat, choosing organic foods may be an option, but it is not an equitable and practical solution for the majority of Americans. Organic foods tend to be more expensive than non-organic products and they are not available for all types of food, stores and areas of the country. Thus, most consumers do not have the genuine choice and access to purchase organic foods as an alternative to what has previously been known as 'traditional' foods (Strauss, 2007). Moreover, issues of cross-contamination increasingly threaten the integrity and economic viability of the organic food supply (WHO, 2005).

The government has the ethical obligation to protect the safety of the mainstream food supply for all of its citizens. The FAO expert panel on ethics recognized that:

[t]he right to adequate food, as understood today, carries with it obligations on the part of states to protect individuals' autonomy and capacity to participate in public decision-making fora, especially when other participants are more powerful, assertive or aggressive. These obligations can include the provision of public resources to ensure that those fora take place in a spirit of fairness and justice. (2001b, p. 25)

The FAO second report concluded that this right has not been fulfilled in connection with genetically engineered products. The most important stakeholders have been excluded from the process because:

[c]itizens have a direct interest in technological developments, yet there are

obstacles to their participation in decision-making that must be acknowledged and overcome. The public has not been adequately informed about the application of gene technology to food production or the consequent potential impacts on consumers' health and the environment. (2001b, p. 25)

As a result, with the confusing and conflicting jumble of claims in the media, 'the public is losing faith in scientists and government' (p. 25).

Following similar reasoning, Geoffrey Podger, as Executive Director of the European Food Safety Authority (EFSA), promoted a labeling approach as a means to regain the support of the public. He explained that the European opposition to GMOs was based on ethical grounds as a reaction to being denied a choice when GMO and non-GMO varieties could not be differentiated. Thus, the European regulatory approach arose in part as a solution to this ethical and practical duty to inform. The advantage of labeling is that it provides a choice '[a]nd while the people who insist on choice may be quite a small part of the population, they are very vociferous and they are often in positions of power and prominence' (Podger, 2004, n.p.). Accordingly, the key to public perceptions is a transparent regulatory process that gives people readily available information on the science.

The responsibility of government to protect its citizens and respond to their concerns should necessitate, at the very least, mandatory labeling and monitoring of GMOs in food. Past studies have found that, unlike the Europeans, American consumers have generally trusted their government and regulatory agencies. Attitudes toward the government link closely with public perceptions of biotechnology, press coverage and policy formation (Gaskell et al., 1999). This fact offers an even greater reason why it is critical that the government does not betray that trust in an area as fundamental and critical as food safety.

Most recently in the USA, public outrage at being denied a choice has generated a grass-roots political effort to raise consciousness of consumers and alert them as to what they are not being told, while advocating labeling (Justlabelit, no date). At the local, state and even federal levels, legislative efforts have attempted to respond to the public's right to know, as well as the safety concerns for consumers and farmers. But a review, below, of the history and current status of those initiatives reveals that these noble goals have not yet been satisfactorily achieved.

Conflicts of interest in studies and scientific research

Several sources have raised the issue of the close connection between the academic community involved in research and the industries or patents they seek to develop (Hoffman and Sung, 2005). This direct financial stake, via stock options or patent participation, creates an inherent conflict of interest. One fear is that 'the lure of profit could color scientific integrity, promoting researchers to withhold information about potentially dangerous side-effects' (Batalion, 2009, n.p.). As a result of this conflict of interest and disincentives for long-term studies, the actual risks of this novel technology remain largely unknown.

Well-funded programs in plant genetics and genetic engineering are supplanting research to enhance organic methods and other low-input alternatives. A 1990 study discovered that 'from [ten percent] up to one third of biomedical researchers at prestigious universities such as Stanford and MIT had direct corporate ties' (Tokar, 1999, n.p.). With the exponential growth of the biotechnology industry since then, today's figures are no doubt even higher. These ties continue to shift more public funds into projects that support the research agenda of the biotechnology industry.

Some groups have also expressed concern that intellectual property incentives limit the development of more beneficial genetically engineered crops. In July 2003, a coalition of public-sector research institutions announced the formation of the Public-Sector Intellectual Property Resource for Agriculture (PIPRA) (Atkinson et al., 2003; PIPRA, no date). Funded by the Rockefeller and McKnight Foundations, PIPRA contends that 'the benefits of much publicly funded research come to private industry through university technology transfer programs, limiting universities' flexibility to conduct research' (Hoffman and Sung, 2005, p. 15). As a result, biotechnology patents may not be utilized for developments with little commercial value that would help the poor and promote the original goal of food security.

Because the research at public institutions is often heavily influenced by the source of funding, this predominantly private backing has diverted research time and money away from projects that would benefit 'the public good, such as biological control, organic production systems and general agroecological techniques' (Altieri and Rosset, 1999, n.p.). This situation has sparked suggestions that '[c]ivil society must request more research on alternatives to biotechnology by universities and other public organizations' (1999, n.p.).

Moreover, one of the significant problems associated with the privatization of biotechnology rights is the company's restriction of information.

> While the basic realities of modern business clearly underscore the need for confidentiality, it is also true that confidentiality provisions are often used as a means of avoiding disclosures. In the face of increasing recognition that activities, including especially species introduction, in one country may have serious impacts on neighbouring countries, labelling and other access to information is increasingly addressed at international and regional levels. (Prakash et al., 2011, p. 7)

Sharing of some of this information is crucial for risk assessment, long-term studies and developing buffer zones between genetically engineered and conventional or organic crops that would avoid spreading these risks exponentially through cross-contamination.

In its recent report, the NAS acknowledged that any new food 'may have some subtle favorable or adverse health effects that are not detected even with careful scrutiny and that health effects can develop over time' (2016, p. 19). It recommended

additional public funding for research, particularly in cases where early published studies produce 'equivocal results' regarding health effects of a GE crop, 'using trusted research protocols, personnel, and publication outlets to decrease uncertainty and increase the legitimacy of regulatory decisions' (p. 19).

According to one examination of the risks and precautions arising from GMOs:

> Regulation of GMO deals with a transscientific problem, that is, the resolution of the problems is beyond the competence of the scientific system. Public perception and acceptance are dependent on trust and whether the products or processes benefit them as citizens and consumers. To take proper accounts of uncertainties and public concern would help to capture the benefits, minimize the risk, and provide goals for future development and use of genetic engineering. (Prakash et al., 2011, p. 11)

Critical to maintaining the public trust will be embracing transparency and disclosure through a comprehensive and effective labeling policy.

The New DARK Act and Preemption of State Initiatives

In the shadow of these ethical tensions emerged a new piece of federal legislation that extinguished grassroots efforts to require mandatory labeling of GMOs in the USA. Attempts to regulate the safety and labeling of GMOs had been raised unsuccessfully every year at the federal level through bills introduced in Congress that died a quiet death in committees. For example, in May 2002, H.R. 4814 was one of five bills introduced by Representative Dennis Kucinich (Democrat-Ohio) that sought to expand the regulation of agricultural biotechnology. On 2 May 2006, Representative Kucinich introduced the 'Genetically Engineered Food Safety Act' (H.R. 5268, 109th Congress 2006) and four other bills regarding GMOs (Strauss, 2007). Representative Kucinich introduced similar bills in the 111th Congress, including legislation that would have pro-tected farmers and shifted liability to the biotech companies (H.R. 5577, 111th Congress 2010) (Strauss, 2012b). One of the most recent efforts, introduced into the US Senate on 2 March 2016 by Democratic Senators Jeff Merkley, Patrick Leahy, Jon Testor and Dianne Feinstein, the 'Biotechnology Food Labeling Uniformity Act' (S. 2621, 114th Congress 2015), would have informed consumers of the presence of GM ingredients in their food while giving several options to food manufacturers for how to indicate this information on the Nutrition Fact Panel (Merkley, 2016). But this proposed legislation was superseded by the one federal bill that did ultimately become law, discussed below.

At the state level, momentum for labeling appeared to build with the passage of several statewide labeling initiatives, which many hoped would take hold and lead to a stringent mandatory labeling under federal law that would follow the states. In 2013, Connecticut became the first state to pass legislation for labeling GMOs in foods, requiring that the retail packaging contain the clear and conspicuous words 'Produced with Genetic Engineering' and redefining the FDA's use of the term 'natural food' to include food that has not been genetically engineered. However, the Connecticut statute provided that the law would not take effect until the passage of similar legislation by four additional Northeast states, with a total aggregate population of more than 20 million, one of which borders Connecticut (Connecticut General Statutes, 2013). Soon after, Maine passed a similar law requiring genetically engineered food and seed stock to be conspicuously labeled as 'Produced with Genetic Engineering' and prohibiting such foods from being labeled as 'natural'. But the Maine statute also contained a 'trigger' clause providing that at least five contiguous states must pass comparable laws (Maine Legislature, 2013). Numerous other states had launched their own labeling bills, some of which were defeated (e.g. California) but were slated to be reintroduced by advocates (Center for Food Safety, no date).

The most successful and definitive of these state laws was passed in Vermont (Vermont General Assembly, 2014) in May 2014, with an accompanying rule, 'Consumer Protection Rule 121', which like the previous states required that food entirely or partially produced with genetic engineering be labeled as such. The law applied to raw agricultural products such as corn and squash as well as processed foods like crackers, soda and cereals. Its comprehensive scheme applied to producers, processors, distributors and retailers. Exemptions included processed foods that would otherwise have been subject to the labeling requirement due to containing one or more materials that have been produced with genetic engineering, in which the genetically engineered materials in the aggregate do not account for more than 0.9% of the total weight of the processed food. The Vermont law was set to go into effect on 1 July 2016, when the biotechnology industry intervened and succeeded in pushing Congress to cut short the state initiatives.

The National Bioengineered Food Disclosure Standard (United States, 2016), pejoratively called the DARK ('Denying Americans the Right to Know') Act like a previous version that had been defeated, was signed into law by President Obama on 29 July 2016. It amended the Agricultural Marketing Act of 1946 to require the Secretary of Agriculture to 'establish a national mandatory bioengineered food disclosure standard with respect to any bioengineered food and any food that may be bioengineered' (Section 293). In a blow to community activists and consumer groups, this statute pre-empted the states from passing their own – and Vermont from implementing its enacted – GMO labeling laws. Furthermore, the USDA sent preemption letters to the governor of every state (USDA, no date(a)). If challenged in court, however, this may be held to be an improper use of preemption, as the federal law currently does not present the requisite characteristics of a comprehensive regulatory scheme. While technically a law requiring labeling, it appears on its face to be so minimal and likely ineffective in its conveyance of information to the consumer that it may be determined to lack adequate and meaningful labeling and thus violate the right of informed consent and effective choice.

With respect to food, the federal statute defines the term 'bioengineering' as a food 'that contains genetic material that has been modified through in vitro recombinant deoxyribonucleic acid (DNA) techniques ... for which the modification could not otherwise be obtained through conventional breeding or found in nature' and limits its scope to foods that are already subject to labeling requirements under the Federal Food, Drug, and Cosmetic Act, the Federal Meat Inspection Act, the Poultry Products Inspection Act and the Egg Products Inspection Act, as well as the predominance of its ingredients (Sections 291 and 292). Although labeling for these substances in food will be mandatory, it gives manufacturers a choice of the disclosure on the package via 'text, symbol, or electronic or digital link' (e.g. QR code accessed with a smartphone) (Section 293(b)(2)(D)). Small food manufacturers can choose to comply instead by placing a telephone number accompanied by appropriate language to indicate that the phone number provides access to additional information and an internet website with disclosure of bioengineering ingredients; the telephone number disclosure must only state, 'Call for more food information' (Sections 293(b)(2)(F)(ii) and 293(d)(1)(B)).

Critics of this new law note that it does not specifically mandate that manufacturers have to post a label or warning that the food contains GMOs. Consumer groups have expressed concern that manufacturers will choose the method that gives the least amount of information or makes it difficult for consumers to ascertain this information in a timely and effective way in order to be able to make a choice before their purchasing decision (Nat, 2016; Begley, 2017). This problem will be exacerbated by technological limitations if the consumer does not have cell service, a phone capable of reading a QR code, or the comfort level and knowledge to do so (Justlabelit, no date; Center for Food Safety, 2017). A national survey confirmed that 88% of American voters prefer GMO labeling printed on the package over bar codes that would be scanned by a smartphone app (Melman Group, 2015). In addition, consumer advocates object to the narrow definition

of 'bioengineered' and scope of the labeling advocated by the industry, urging instead that 'all foods produced through genetic engineering are labeled; including those derived from genetically engineered sources, such as highly refined sugars and oils and processed corn and soy ingredients' (McCann, 2017, n.p.). Moreover, any threshold set by the USDA should be consistent with international standards – the mandatory disclosure of 0.9% by individual GE ingredient (2017, n.p.).

Unlike previous statutes that put food labeling in the purview of the FDA, the law designated the Agricultural Marketing Service (AMS) of the USDA as the agency responsible for promulgating additional regulation that would fill in the details to establish a 'national mandatory system for disclosing the presence of bioengineered material' (USDA, no date(a)). The USDA has formed a working group to develop a timeline for rule-making and posted questions on its website in the summer of 2017 to seek input on a series of issues left open by the legislation to aid in its drafting of the rules. These questions include the scope, threshold levels and definitions of a 'bioengineered food'; the type of text, symbol or digital/electronic link that should be designated 'if a manufacturer chooses to use' it; alternatives for 'very small or small packages'; and definitions for the stated exclusions, such as 'small' and 'very small manufacturers', 'restaurants and similar retail food establishments' (USDA, no date(b)). AMS received over 112,000 responses from individuals and organizations to be used in its drafting of the proposed rule, issued with an additional public comment period.

The USDA also commissioned a study 'to identify potential technological challenges that may impact whether consumers would have access to the bioengineering disclosure through electronic or digital disclosure methods' and published the results on 7 September 2017 (USDA, 2017b). Among its most significant findings was the fact that 85% of consumers experienced technical challenges when using mobile apps for scanning digital links: 'In addition, most apps contain advertisements that confuse consumers and run counter to how the Law requires disclosure when regulations are finalized and implemented' (2017b, p. 4). Pursuant to the law, the USDA is required to have the final rules in place by 29 July 2018, two years after the passage of the legislation.

In view of prospective future court challenges, it would be prudent for the USDA to take the opportunity to strengthen labeling requirements to a level akin to Vermont and other state initiatives or international standards. The Vermont Attorney General's Office has already announced its intention to 'take an active role as the labeling fight shifts from the legislative process in Congress to the regulatory process at the USDA ... We will work hard to give consumers the same access to information, in plain English, that they had under Vermont's law' (2016, n.p.). Pressure will be wielded by consumers, anti-GMO groups and other stakeholders to determine if this will be only the first step in a comprehensive and effective mandatory federal labeling scheme or a temporary defeat for states that will ultimately prevail over federal obstructionism. One way or another, the US government must respond to the increasing crescendo of the public outcry for transparency and disclosure in the debate over food safety. Hopefully these governmental agencies will be guided by their charge and uphold their duty to the public to take into account consumer concerns regarding labeling and to make paramount the safety of the food supply for the average American consumer.

A Framework that Addresses these Ethical Tensions

In light of ethical concerns and individuals' comfort levels in assessing whether the risks are acceptable, American consumers should not be bound to accept the possible consequences of GM foods without their knowledge and consent. With scientific uncertainty about the risks of consuming GM foods unlikely to be resolved in the near future, ethical concerns should be the paramount factor in determining a model of labeling, segregating and monitoring for GM foods.

Reflecting on the ethical tensions that biotechnology creates prompts one to consider questions that should be raised and utilized to construct a more appropriate regulatory response. The current regulatory framework is inadequate because it predates the advent of this technology and forces regulators to conform substances created through a novel process into preexisting and longstanding notions of safety and efficacy. Better public policy would result from approaching this area from an alternative perspective. Under what conditions do consumers have a right to know the process by which their food has been produced? The US already requires labeling to inform consumers about some food production cases (e.g. irradiation). Given the uniqueness of biotechnology and its potential for harm, should this change the way we think about the right to know and implore us to answer this question in the affirmative?

The way of approaching market approvals and labeling requirements in the past has been to focus only on the composition and safety of the product itself without considering the potential risk inherent in the technology. Biotechnology is so fundamentally different that it forces us to rethink that approach. Biotechnology raises issues separate from whether to label the product as to the presence of a particular ingredient or lack thereof, which has been the traditional approach of the FDA. Genetic engineering of food also goes beyond the existing criteria for labeling of the potential for harm embodied by the concept of 'substantial equivalence', regardless of whether the product itself is or is not different. Under the current FDA policy, if a genetically engineered product does not appear to be substantially different than its conventional counterpart, no special labeling or animal testing is required. 'From FDA's perspective, biotechnologically-produced products are seen as substantially equivalent to conventional food products because, in the agency's view, there is no scientific basis to presuppose that biotech foods are more risky or substantially different from other food products' (Strauss, 2006, p. 183). However, there is no definition provided in the regulations for substantial equivalence or clear guidelines stipulating what to examine and how similar the items in question should be; this concept has been discredited in Europe as unscientific and not adequate to justify a lack of safety assessments (Strauss, 2006). Biotechnology is so extraordinary that there are many unknowns and scientific uncertainty, with potential risks inherent in the technology itself. This innovation calls for a new way of thinking because it would be inappropriate for regulators to impose an existing framework on it. Policy makers should recognize this matter as unique due to the unknowns and possible risks.

The issue of risk is made even more compelling due to the fact that science has not yet advanced to measure and assess these risks using standard risk assessment techniques. Thus policy makers cannot turn to science alone to provide the answers to the pressing questions. A more precautious policy could create the proper incentives for the development of more advanced risk assessment procedures, as well as improved detection methods and long-term studies of safety. Moreover, a more prudent policy with greater emphasis on public safety concerns would be consistent with the government's recent proactive approach towards food safety embodied in the Food Safety Modernization Act (Strauss, 2011). Until such time as these risks can be ascertained with a higher degree of certainty, consumers of these products should be given a choice and allowed to weigh their own personal comfort levels in their purchase and consumption decisions. In view of this scientific uncertainty, effective labeling should be an important component of the revised regulatory framework, along with an education campaign as to the meaning and significance of these GE designations.

This discussion relates to a more fundamental question: when a new technology is introduced, when do people have the right to be informed and what do they have a right to be informed about (e.g. that the product was made through this method)? This concern goes beyond a focus on the potential for harm. As the Terminator gene illustrates, there is no limit to what might be developed

through genetic engineering because biotechnology is so profoundly different from traditional crop and animal breeding practices. As a consequence, society needs to develop a different way of looking at it – at the very least labeling and educating the public – and ultimately introducing a new regulatory scheme so that as the science and technology develops, new inventions will be incorporated into that scheme. For example, requiring better detection methods and risk assessment for the approval process would both provide incentives to develop that new technology and guide the future direction of the science.

A consideration of these issues highlights the fact that ethical principles must shape the solution for the treatment of biotechnology food products. The incorporation of ethics into policy development should involve all stakeholders: farmers, consumers, the environment, underprivileged populations and the agricultural biotechnology industry. Recognition of the inherent conflicts of interest also necessitates a more active and independent role of regulatory agencies in relation to the biotechnology companies. An important part of this framework would be clear and comprehensive labeling, which is essential for informed consumer choice, as well as a sharing of information and education on the science, including all potential and discovered risks to human health and the environment.

A group of scientific experts identified three important components for risk management: impact assessment, public awareness/participation and the design of regulatory systems. Above all is involvement of the public: 'It is not possible to overstate the importance of the public's contribution to effective decision making, as well as the importance of public awareness, within the context of government decisions on matters and activities affecting the environment' (Prakash et al., 2011, p. 6). Under this analysis, disclosure and transparency should extend beyond public access to governmental documents and processes:

> [T]here are other mechanisms by which public awareness and access to information can be encouraged, including product

labeling, food safety standards, and general consumer protection laws, all of which are designed to foster awareness and communicate public preferences to the commercial proponents of GMOs in a way that will get their attention. (2011, p. 6)

In order to be effective, such labeling must be 'accurate, specific, and clearly expressed in understandable language, unbiased, and based on full disclosure of the relevant facts by the GMO proponents' (Prakash *et al.*, 2011, p. 6).

The WHO study also recognized the need and responsibility for communicating risks to the public so that 'ethical components of food-safety decisions are clearly identified as early in the process as possible' and 'value-laden choices made by risk managers are made in an open, participatory process that respects the rights and roles of all stakeholders' (2005, p. 56).

Conclusion

In view of the most recent federal labeling statute in this area, the predominant query to consider may be the following: Do the current regulations, or lack of adequate and meaningful labeling, violate our responsibilities to others by not allowing them a choice as to whether they knowingly and willingly assume the risks of ingesting these genetically engineered substances? Efforts to quell the public outcry by preempting state and local community labeling initiatives would be better directed to incorporating these desires into the international policy arena.

The ethical implications are clear, followed by the expectation that the legal system should fill in the ethical gap as it has done in so many other areas and, at the very least, require meaningful labeling, pre-market approval and monitoring of GMOs in food products and ingredients. Fully informing the public and transparency in the regulatory process are the keys. Incorporating ethical tensions in the policy and regulation arena requires US policy makers to respond appropriately to the consumer demands that led to grassroots efforts and embrace rather

than suppress these initiatives. Mandatory labeling that clearly and effectively discloses genetically engineered ingredients and processes would be the responsible next step.

References

Altieri, M.A. and Rosset, P. (1999) Ten reasons why biotechnology will not ensure food security, protect the environment and reduce poverty in the developing world. *AgBioForum* 2, 3. http://www.agbioforum.org/v2n34/v2n34a03-altieri.htm

Atkinson, R.C. *et al.* (2003) Public sector collaboration for agricultural IP management. *Science* 301, 174–175. doi:10.1126/science.1085553

Batalion, N. (2009) 50 harmful effects of genetically modified food. Available at: http://www.raw-wisdom.com/50harmful (accessed 12 August 2017).

Begley, C. (2017) 'So close, yet so far': The United States follows the lead of the European Union in mandating GMO labeling. But did it go far enough? *Fordham International Law Journal* 40, 625–746. Available at: http://heinonline.org/HOL/Landing Page?handle=hein.journals/frdint40&div=17 (accessed 5 July 2018).

Brussel, L. (2003) Engineering a solution to market failure: a disclosure regime for genetically modified organisms. *Cumberland Law Review* 34, 427. Available at: http://heinonline.org/HOL/LandingPage?handle=hein.journals/cumlr34&div=25 (accessed 5 July 2018).

Center for Food Safety (n.d.) State bills. Available at: http://www.centerforfoodsafety.org/search/state-bills (accessed 23 July 2017).

Center for Food Safety (2017) U.S. polls on GE food labeling. Available at: http://www.centerforfoodsafety.org/issues/976/gefood-labeling/us-polls-on-ge-food-labeling# (accessed 21 July 2017).

Connecticut General Statutes (2013) An Act Concerning Genetically-Engineered Food (Public Act No. 13-183). Available at: https://www.cga.ct.gov/2013/ACT/pa/pdf/2013PA-00183-R00HB-06527-PA.pdf (accessed 23 July 2017).

Consumer Reports (2014) Where GMOs hide in your food. October. Available at: http://www.consumerreports.org/cro/2014/10/where-gmos-hide-in-your-food/index.htm (accessed 21 July 2017).

Convention on Biological Diversity (2017) Genetic use restriction technologies (GURTs). Available at: https://www.cbd.int/agro/gurts.shtml (accessed 21 July 2017).

FAO (2001a) Ethical issues in food and agriculture. Available at: http://www.fao.org/DOCREP/003/X9601E/X9601E00.HTM (accessed 1 December 2017).

FAO (2001b) Genetically modified organisms, consumers, food safety and the environment. Available at: http://www.fao.org/3/a.x9602e.pdf. (accessed 1 December 2017).

FAO (2005) Report of the panel of eminent experts on ethics in food and agriculture, Third session. Available at: http://www.fao.org/docrep/010/a0697e/a0697e00.html (accessed 1 December 2017).

FAO (2007a) Panel of eminent experts on food and agriculture. Available at: http://www.fao.org/unfao/govbodies/gsb-subject-matter/statutory-bodies-details/en/c/432/?no_cache=1 (accessed 22 July 2017).

FAO (2007b) Report of the panel of eminent experts on ethics in food and agriculture, Fourth session. Available at: http://www.fao.org/docrep/014/i2043e/i2043e.pdf (accessed 1 December 2017).

FDA (2015) Guidance for industry: voluntary labeling indicating whether foods have or have not been derived from genetically engineered plants. Available at: https://www.fda.gov/Food/GuidanceRegulation/GuidanceDocumentsRegulatoryInformation/ucm059098.htm (accessed 21 July 2017).

FDA (2016) Food irradiation: What you need to know, 28 June. Available at: https://www.fda.gov/food/resourcesforyou/consumers/ucm261680.htm (accessed 27 July 2017).

Gaskell, G., Bauer, M.W., Durant, J. and Allum, N.C. (1999) Worlds apart? The reception of genetically modified foods in Europe and the U.S. Science 285, 384–387. doi:10.1126/science.285.5426.384

GMA (2017) Grocery Manufacturers Association position on GMOs. Available at: http://www.gmaonline.org/file-manager/GMA%20Position%20on%20GMOs.pdf (accessed 21 July 2017).

Halloran, J. and Hansen, M. (1999) Why we need labeling of genetically engineered food. Synthesis/Regeneration 18. Available at: http://www.greens.org/s-r/18/18-07.html (accessed 27 July 2017).

Hoffman, D.E. and Sung, F. (2005) Future public policy and ethical issues facing the agricultural and microbial genomics sectors of the biotechnology industry. Biotechnology Law Report 24, 10–28. doi:10.1089/blr.2005.24.10

Justlabelit (n.d.) GMO labeling in Congress. Available at: http://www.justlabelit.org/dark-act/ (accessed 23 July 2017).

Keane, S. (2006) Can a consumer's right to know survive the WTO? The case of food labeling. Transnational Law and Contemporary Problems 16, 291.

Kolehmainen, S. (2001) Precaution before profits: An overview of issues in genetically engineered food and crops. Virginia Environmental Law Journal 20, 267–294. Available at: http://www.jstor.org/stable/24785929 (accessed 5 July 2018).

Maine Legislature (2013) Maine revised statutes, title 22, § 2591. Available at: http://legislature.maine.gov/legis/statutes/22/title22sec2591.html (accessed 23 July 2017).

McCann, N. (2017) Now is our chance to get USDA to label GMOs on the package, 13 July. Available at: http://www.centerforfoodsafety.org/blog/5016/now-is-our-chance-to-get-usda-to-label-gmos-on-the-package# (accessed 23 July 2017).

Melman Group (2015) Voters want GMO food labels printed on packaging. Available at: http://4bgr3aepis44c9bxt1ulxsyq.wpengine.netdna-cdn.com/wp-content/uploads/2015/12/15memn20-JLI-d6.pdf (accessed 23 July 2017).

Merkley, J. (2016) Press Release, Merkley, Leahy, Tester, Feinstein introduce GMO food labeling legislation, 2 March. Available at: https://www.merkley.senate.gov/news/press-releases/merkley-leahy-tester-feinstein-introduce-gmo-food-labeling-legislation (accessed 23 July 2017).

Nat, H. (2016) Will consumers be in the 'dark' about labels on genetically engineered and modified foods? Journal of Food Law & Policy 12, 199–213. Available at: http://heinonline.org/HOL/Landing Page?handle=hein.journals/jfool12&div=13 (accessed 5 July 2018).

National Academies of Sciences, Engineering, and Medicine (NAS) (2016) Genetically Engineered Crops: Experiences and Prospects. NAS, Washington, DC. doi:10.17226/23395

Nauheim, D.A. (2009) Food labeling and the consumer right to know: Give the people what they want. Liberty University Law Review 4, 97–133. Available at: http://heinonline.org/HOL/Landing Page?handle=hein.journals/lunlr4&div=5 (accessed 5 July 2018).

Nestle, M. (1996) Allergies to transgenic foods: Questions of policy. New England Journal of Medicine 334, 726–728. doi:10.1056/NEJM199603143341111

Nordlee, J.A. et al. (1996) Identification of a Brazil nut allergen in transgenic soybeans. New England Journal of Medicine 334, 688–692. doi:10.1056/NEJM199603143341103

PIPRA (n.d.) Public intellectual property resource for agriculture. Available at: http://www.pipra.org (accessed 28 July 2017).

Podger, G. (2004) European food safety authority will focus on science. The European Institute V(1). Available at: http://www.europeaninstitute.org/index.php/ei-blog/32-european-affairs/winter-2004/309-european-food-safety-authority-will-focus-on-science (accessed 28 July 2017).

Prakash, D., Verma, S., Bhatia, R. and Tiwary, B.N. (2011) Risks and precautions of genetically modified organisms. *ISRN Ecology* 2011, 369573. doi:10.5402/2011/369573

Strauss, D.M. (1987) Reaffirming the Delaney Anti-cancer Clause: The legal and policy implications of an administratively created *de minimis* exception. *Food Drug Cosmetic Law Journal* 42, 393–428. Available at: http://heinonline.org/HOL/LandingPage?handle=hein.journals/foodlj42&div=32 (accessed 5 July 2018).

Strauss, D.M. (2006) The international regulation of genetically modified organisms: Importing caution into the U.S. food supply. *Food and Drug Law Journal* 61, 167–168. Available at: http://heinonline.org/HOL/LandingPage?handle=hein.journals/foodlj61&div=17 (accessed 5 July 2018).

Strauss, D.M. (2007) Defying nature: The ethical implications of genetically modified plants. *Journal of Food Law & Policy* 3, 1–37. Available at: http://heinonline.org/HOL/LandingPage?handle=hein.journals/jfool3&div=5 (accessed 5 July 2018).

Strauss, D.M. (2009) The application of TRIPS to GMOs: International intellectual property rights and biotechnology. *Stanford Journal of International Law* 45, 287–320. Available at: https://ssrn.com/abstract=1523514 (accessed 5 July 2018).

Strauss, D.M. (2011) An analysis of the FDA Food Safety Modernization Act: Protection for consumers and boon for business. *Food and Drug Law Journal* 66, 353–376. Available at: https://ssrn.com/abstract=1925008 (accessed 5 July 2018).

Strauss, D.M. (2012a) Liability for genetically modified food: Are GMOs a tort waiting to happen? *The SciTech Lawyer* 9, 8–13. Available at: https://ssrn.com/abstract=2162255 (accessed 5 July 2018).

Strauss, D.M. (2012b) The role of courts, agencies, and Congress in GMOs: A multilateral approach to ensuring the safety of the food supply. *Idaho Law Review* 48, 267–319. Available at: http://heinonline.org/HOL/LandingPage?handle=hein.journals/idlr48&div=14 (accessed 5 July 2018).

Teitel, M. and Wilson, K. (1999) *What You Need to Know to Protect Yourself, Your Family, and Your Planet*. Park Street Press, Rochester, Vermont.

Tokar, B. (1999) Resisting biotechnology and the commodification of life. *Synthesis/Regeneration* 18. Available at: http://www.greens.org/s-r/18/18-01.html (accessed 26 July 2017).

United States (2012) Irradiation in the production, processing, and handling of food, Final Rule. *Federal Register* 77, 71316–71321. Available at: https://www.federalregister.gov/documents/2012/11/30/2012-28968/irradiation-in-the-production-processing-and-handling-of-food (accessed 1 December 2017).

United States (2016) National bioengineered food disclosure standard, amending the Agricultural Marketing Act of 1946, Public Law 114-216 (7 U.S.C. 1621 et seq.) (2012). Available at: https://www.congress.gov/114/plaws/publ216/PLAW-114publ216.pdf (accessed 22 July 2017).

USDA (n.d.a) GMO disclosure and labeling. Available at: https://www.ams.usda.gov/rules-regulations/gmo (accessed 9 December 2017).

USDA (n.d.b) Proposed rules questions under consideration. Available at: https://www.ams.usda.gov/rules-regulations/gmo-questions (accessed 9 December 2017).

USDA (2017a) Adoption of genetically engineered crops in the U.S., 12 July. Available at: https://www.ers.usda.gov/data-products/adoption-of-genetically-engineered-crops-in-the-us/ (accessed 21 July 2017).

USDA (2017b) Study of electronic or digital disclosure, 6 September. Available at: https://www.ams.usda.gov/reports/study-electronic-or-digital-disclosure; https://www.ams.usda.gov/sites/default/files/media/USDADeloitteStudyofElectronicorDigitalDisclosure20170801.pdf (accessed 9 December 2017).

Vermont Attorney General's Office (2016) Press Release, Attorney General will not enforce GE food labeling law, will advocate for clear on-package labels, 2 August. Available at: http://ago.vermont.gov/focus/news/attorney-general-will-not-enforce-ge-food-labeling-law-will-advocate-for-clear-on-package-labels.php (accessed 23 July 2017).

Vermont General Assembly (2014) *H.112 (Act 120): An act relating to the labeling of food produced with genetic engineering*. Available at: http://www.leg.state.vt.us/docs/2014/Acts/ACT120.pdf (accessed 1 December 2017).

World Health Organization (WHO) (2005) Modern food biotechnology, human health and development: An evidence-based study. Available at: http://www.who.int/foodsafety/publications/biotech/biotech_en.pdf (accessed 24 July 2017).

7

Ethical Tensions in Regulation of Agricultural Biotechnology and their Impact on Policy Outcomes: Evidence from the USA and India

Deepthi E. Kolady[1]* and Shivendra Kumar Srivastava[2]

[1]*Department of Economics, South Dakota State University, Brookings, South Dakota, USA;* [2]*ICAR-National Institute of Agricultural Economics and Policy Research, New Delhi, India*

Introduction

Currently, genetically engineered (GE) crops are grown in 28 countries by about 18 million farmers over 179.7 million hectares. Globally, 83% of soybean area, 75% of cotton area, 29% of maize area and 24% of canola area are planted with GE varieties (James, 2015). In a recent report analyzing the available evidence on the impacts of adoption of GE crops, the US National Academy of Sciences, Engineering, and Medicine (NAS) concluded that GE crops generally had favorable economic outcomes for producers of all scales who have adopted these crops, but outcomes have been heterogeneous depending on pest abundance, farming practices and agricultural infrastructure, and the foods from GE crops are as safe as the foods from non-GE crops (NAS, 2016).

Despite the availability of such scientific consensus on the positive impacts, there are cross-national differences in the way governing bodies and policy makers have reacted to the use of GE technology, across the globe. For example, the USA has a more permissive regulatory policy and the European Union (EU) has a more stringent policy process (Bernauer and Meins, 2003; Graff *et al.*, 2009; Cowan, 2015). However, India, the regional leader in the research and development

(R&D) of agricultural biotechnology, sends mixed signals on the regulation and use of GE crops. For example, the Minister of Environment, Forests & Climate Change in India had overridden the recommendation of the scientific body to commercialize its first GE food crop Bt eggplant and imposed a moratorium on commercialization. Similarly, despite the recommendation by the scientific body to commercialize its first public-sector developed GE crop, hybrid mustard, India is currently witnessing an intense debate on its commercialization (Chauthaiwale, 2016; Aggarwal, 2016).

Previous studies have used political economy theories which take into account the influence that groups of regulated economic agents have on regulators to explain the cross-national differences on policy and regulation of agricultural biotechnologies. Capture theory proposed by Posner (1974) suggests that the coalition of biotech industry and export-oriented farmers captures regulators in the USA, while the coalition of environmental activists, consumer groups and green party politicians captures regulators in the EU. Zusman (1976) proposed a political economy model where the regulator weighs the interests of various groups within the economy to incorporate and measure social power in economic systems

* Corresponding author. E-mail: deepthi.kolady@sdstate.edu

and policy-making. Building on the framework of Zusman (1976), Graff *et al.* (2009) proposed a nuanced three-stage theory where interactions among competing interests are allowed to explain the political economy behind regulation of agricultural biotechnologies. Another popular argument, building on the interest group theory, is that the differences in consumers' underlying attitudes and perceptions towards risk account for the regulatory policy differences between the EU and the USA (Bernauer and Meins, 2003). However, most of these studies examined how competing interest groups influence regulatory outcomes for GE crops in developed countries.

Although most of the currently available GE traits focus on economically important crops and traits in developed countries, the potential of GE technology to address food and nutrition insecurity in developing countries is huge. Additionally, most of the GE technologies currently used in developing countries have their origins in the USA, except for China and for some cases in India. However, for a variety of scientific, economic, social and regulatory reasons, many GE traits and crop varieties that have been developed are not in commercial production in many countries, including the USA, the EU and India. This suggests that formation of policies for GE crops involve factors other than technical risk assessments and raises the question whether there are ethical concerns relating to GE crops' regulatory policy that create a conflict of interests, values or rights between various stakeholders. We are unaware of any study that analyzes why and how ethical tensions arise because of agriculture biotechnology policy and whether there are any cross-national differences in the way these ethical tensions affect regulatory policy frameworks and outcomes. To our knowledge, studies analyzing the drivers of regulatory differences between developed and developing countries are limited. In this chapter we address these gaps in the literature by using the USA and India as case examples to analyze how ethical tensions arise related to agriculture biotechnology policy and how institutional framework and domestic politics

address these ethical tensions and contribute to divergence in regulatory policy outcomes.

Ethical Concerns and Ethical Tensions

Development of a regulatory framework for agricultural biotechnology and its products began with the voluntary recommendations of the 1975 Asilomar meeting, the first International Congress on Recombinant DNA molecules in California, where the scientific community discussed the potential benefits and risks of this technology (Berg, 2008). The USA took the initiative to implement the Asilomar recommendations, which were later followed up by other countries and international organizations such as the Food and Agricultural Organization (FAO) and World Health Organization (FAO/WHO, 2015).

The regulatory policies are formulated and enforced to promote social welfare enhancing technologies while ensuring the safety of human, animal and environmental health. This objective itself creates the potential for conflicts in interests, values and rights of various stakeholders of GE crops, including technology developers, scientists, farmers, consumers, food processing firms, retail firms and civil society within a country. Many of the expressed concerns with GE crops and foods are described as ethical, which means that these concerns demand for competing visions of nature and public good and some stakeholders feel that their values or rights are neglected and demand for their voices to be heard. Thompson (2000) categorized these ethical concerns into three broad types: (i) special concerns related to the use of agriculture biotechnology; (ii) general concerns related to any technology; and (iii) concerns on regulators' responses to the problems of technological ethics.

The arguments of species boundaries, genes and essence, natural kinds, and religion contribute to special ethical concerns (Thompson, 2000). Critics of GE technology can use these arguments to oppose commercialization of GE technology, which can lead to unique and focused ethical tensions between stakeholders. Examples of such ethical tensions include those between religious groups and users of GE technology. Although currently not dominant, special ethical concerns show that modern biotechnology is challenging the ways in which humans have made sense of the world since its origins. If domestic politics or religious institutions push these special concerns to the forefront of policy discourse, they can gain widespread appeal and create ethical tensions which can potentially stall the regulatory process indefinitely. While regulators cannot address these concerns through policy formulations, public education and outreach are the best possible ways to address these special ethical concerns.

The arguments of food safety, environmental health and safety impact on farming communities and farm structure, and shifting power relations among and between various stakeholders contribute to general concerns related to technology use in agriculture (Thompson, 2000). These concerns are considered general because they are not specific to agriculture biotechnology, rather they are applicable to any technology used in agriculture. Additionally, these concerns lead to ethical tensions between a broad category of stakeholders, including scientists, farmers, private firms, consumers, environmentalists and civil society. Ethical tensions related to food safety issues arise from concerns related to the likelihood of potential for hazards and harm from consuming agriculture biotechnology derived products. Ethically significant environmental impacts of technology include human health effects due to environmental exposure, catastrophic ecological consequences which could destabilize global food systems and threaten human society, and non-anthropocentric environmental impacts felt less by humans and more by the ecosystem. Concerns such as that agriculture biotechnology will contribute to narrowing of genetic diversity in major food crops and development of insect resistance or super weeds fall under the category of ecological consequences with potential catastrophic effects. Concerns related to the influence of the private sector on seed industry

structure and seed prices, farm structure and food sovereignty, farmers' access to GE seeds, impact of GE seed adoption on farmers' net incomes and liability issues related to the failure of GE seeds come under the broader category of social justice-related ethical concerns. These concerns lead to ethical tensions between various stakeholders, particularly between private firms, farmer groups, environmentalists, consumer groups and civil society.

One possible approach to address general ethical concerns related to technology use is to base regulatory decision on the best available scientific evidence and follow the ethical logic of informed consent. Regulatory frameworks in almost all countries include technical risk assessments to address the concerns related to environmental health and food safety. However, responses from regulators and technology developers have created further ethical concerns. As we show later in the chapter, there are increasing concerns about the credibility of environmental impact assessments and regulators' technical and financial ability to conduct such assessments. Ethical concerns associated with regulators' responses to technology ethics also involve questions such as labeling and consumer choice. Figure 7.1 shows the pathways in which ethical concerns affect regulatory policies and outcomes.

Regulatory System in the USA

In the early 1980s, the US government agencies struggled with questions such as whether to regulate agricultural biotechnology and whether to adopt process- or product-oriented approaches to regulation. While the White House Office of Science and Technology Policy, the United States Department of Agriculture (USDA) and the Federal Drug Administration (FDA) preferred promotion of biotechnology and a more permissive product-oriented approach, the United States Environmental Protection Agency (USEPA) advocated the development of new risk assessment procedures and a process-based approach to regulation. In 1984, the USEPA

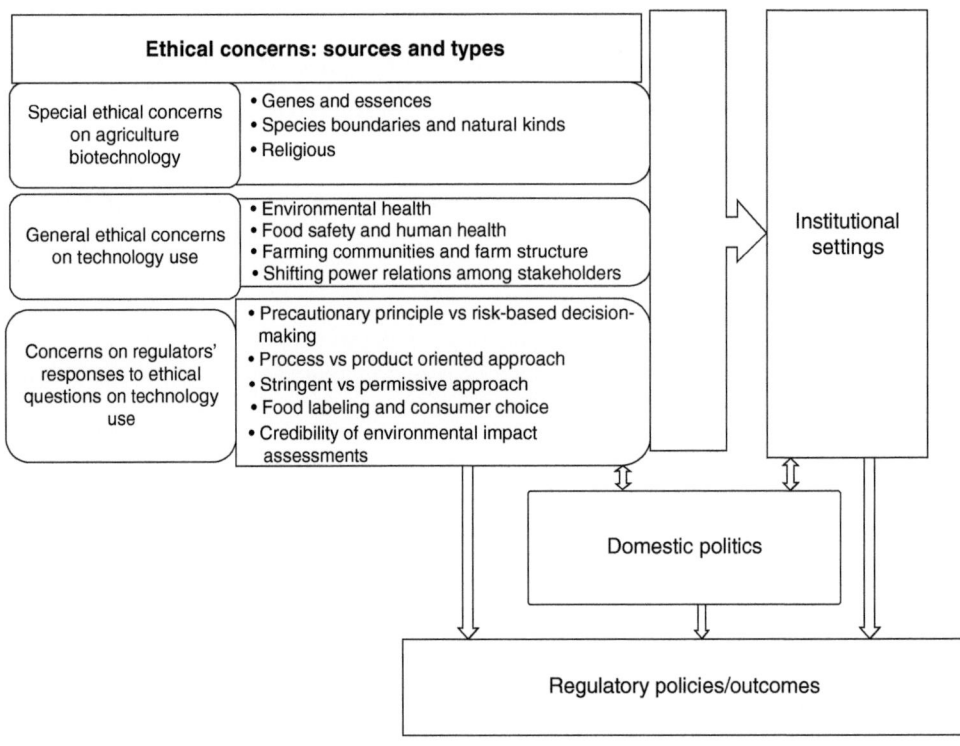

Fig. 7.1. Pathways in which ethical concerns affect regulatory policies and outcomes. (Deepthi E. Kolady and Shivendra Kumar Srivastava.)

proposed that the regulation of biotechnology should be under the Federal Insecticide, Fungicide, and Rodenticide Act and the Toxic Substance Control Act (Bernauer and Meins, 2003). To address the criticism of lack of coordination in biotechnology regulatory policy, the Reagan Administration created a Cabinet Council Working Group in 1984, with representatives from 15 agencies. In 1986, the Working Group issued a Coordinated Framework for the Regulation of Agricultural Biotechnology, which established the product orientation of regulatory policy and denied the demand for new legislation specifically for genetically engineered organisms. The key principle behind the Coordinated Framework is that biotechnology processes do not pose any special risks, hence it demands no new laws beyond those that already govern the health, safety, efficacy and environmental impact of traditional production methods. According to the Coordinated Framework, it is the product that is regulated,

not the process. Thus the Coordinated Framework adopted the *substantial equivalence* approach that measures whether or not GE crops or foods show more variation in measured health and nutritional characteristics as compared to samples of non-GE conventional counterparts (Hochman *et al.*, 2008). The Coordinated Framework assigned the primary responsibility for regulation of agricultural biotechnology to the USDA, the FDA and the USEPA, and established principles for coordination and cooperation among these authorities. As we show below, the federal agencies responsible for regulation try to address many of the general ethical concerns we discussed earlier (Table 7.1).

The Animal and Plant Health Inspection System (APHIS) is part of the USDA and regulates the importation, inter-state movement and field testing of GE plants and organisms that are or might be plant pests under the Plant Protection Act. The APHIS also regulates animal biologics and approves

Table 7.1. Institutional framework for regulation of agriculture biotechnology in the USA. (Deepthi E. Kolady and Shivendra Kumar Srivastava, various sources.)

Category	APHIS (USDA)	EPA (USEPA)	FDA (USDHHS)
Ethical concern addressed	Environmental, human and animal health	Environmental and human health	Human and animal health and food safety
Focus	Biotechnology-derived plants, veterinary biologics	Plant-incorporated protectants, microbial pesticides	Food, drugs, medical devices, animal drugs
Role	Approve import, interstate movement, field testing and environmental release	Register new pesticides, decide the tolerance levels of pesticide residues in raw and final products, experimental use permit (EUP), field testing	Enforce safety of food and drug supplies, decide on food labeling issues and safety of food additives, mandatory pre-market consultations since 2001
Provision for public comment	Yes	Yes	Yes
Decision time	10 days for request on field tests & interstate movement, 30 days for request on environmental release, 180 days for request on non-regulated status	90 days for pre-manufacturer application	120 days for pre-market biotechnology notice

environmental release of GE products. The agency guides testing procedures to prevent escape of regulated products and to avoid any harm to agriculture, human health or the environment. Regardless of the process chosen, the developer seeks 'non-regulated status' for commercialization after the testing is completed. Thus the APHIS addresses general ethical concerns related to environmental, animal and human health.

The USEPA registers and approves the use of all pesticides. It considers pesticides generally engineered into plants as 'plant-incorporated protectants'. By deciding the tolerance levels of pesticide residues in raw and final products of plant-incorporated protectants such as Bt corn/soybean and by issuing experimental use permits for manufacturing and field testing, the USEPA addresses general ethical concerns related to environmental and human health.

The FDA is part of the US Department of Health and Human Services (USDHHS). It regulates food, animal feed additives, and human and animal drugs to ensure that they pose no human health risks under the Federal Food, Drug, and Cosmetic Act and the Public Health Service Act. Because of the substantial equivalence approach, GE-derived foods have to be regulated like non-GE foods if their nutritional composition does not differ from their conventional counterparts. By ensuring the safety of food and drug supplies and food additives, the FDA addresses the general ethical concerns related to food safety and human and animal health.

Regulatory System in India

Encouraged by the scientific breakthroughs in the field of biotechnology in the 1980s, the Government of India established the National Biotechnology Board (NBB) in 1982. Given India's experience with the Green Revolution, which reduced the impact of famines within the country, it is not surprising that there was early recognition of the potential of biotechnology and its acceptance as a tool to address food security issues. The NBB drafted and issued biosafety guidelines in 1983 to undertake biotech-related research in laboratories and contained use settings. The NBB later became the Department of Biotechnology (DBT) under the Ministry of Science and Technology (DBT, 2016). The DBT is the federal agency promoting biotechnology-related research in the country. Since the Ministry of Environment, Forests & Climate Change (MOEF&CC) is responsible for biodiversity conservation and environmental protection in the country, the MOEF&CC started regulating GE crops and their products under the existing Environmental Protection Act (EPA), commonly referred to as EPA 1986. Since EPA 1986 does not deal with GE organisms or products *per se*, the MOEF&CC issued the 'Rules for manufacture, use, import, export, and storage of hazardous microorganisms, genetically engineered organisms or cells', known as 'EPA Rules 1989' (DBT, 2016). According to the MOEF&CC, all GE products are regulated in view of potential risks to human and environmental health. The EPA Rules 1989 have proposed six competent statutory authorities in India (Table 7.2).

The Recombinant DNA Advisory Committee (rDAC) functions in an advisory capacity to the food and technology industries and to the public sector. It reviews developments in biotechnology at national and international levels and recommends appropriate safety regulations in recombinant research, use and applications. The federal ministry of Science and Technology administers the rDAC. By recommending safety regulations for recombinant research, use and applications, the rDAC addresses ethical concerns related to biosafety, which include human and animal health and food safety. Any institution engaged in recombinant research can establish and administer an Institutional Biosafety Committee (IBSC) to oversee such research at the institutional level. By ensuring that each and every institution engaged in biotechnology research follows the biosafety protocols, the IBSC addresses general ethical concerns related to biosafety.

The Review Committee on Genetic Manipulation (RCGM) monitors the safety-related aspects of ongoing research projects and activities. The RCGM produces manuals and guidelines specifying procedures for regulatory process with respect to activities involving GE organisms in research, use and applications, including industry, with a view to ensure environmental safety. The federal ministry of Science and Technology administers the RCGM. The Genetic Engineering Appraisal Committee (GEAC) is the apex body that grants approval for activities involving large-scale use of hazardous microorganisms and recombinants in research and industrial production. It is also responsible for approval of proposals relating to release of GE organisms and products into the environment, including experimental field trials. The federal ministry of Environment, Forests and Climate Change administers this. Both the RCGM and GEAC address general ethical concerns related to biosafety and environmental health.

The State Biotechnology Coordination Committee (SBCC) has a major role in monitoring, including powers to inspect, investigate and take punitive action in case of violations of statutory provisions. Respective state governments administer the SBCC. The District Level Committee (DLC) has a major role in monitoring the safety regulations in installations engaged in the use of GE organisms/hazardous microorganisms and their applications in the environment. Respective state governments administer the DLC. Both the SBCC and DLC have supervisory roles and address ethical concerns related to biosafety and environmental health, particularly in the post-release stage.

This brief discussion shows that the institutional framework in India addresses general ethical concerns related to biosafety and environmental health. However, it is important to note that the final decision on environmental release of GE organisms for commercial use is taken not by the apex regulatory authority, the GEAC, but by the MOEF&CC. Additionally, the environment is the responsibility of the federal government and agriculture is the responsibility of state governments in India. Thus, regulations for agricultural biotechnology products primarily come under the purview of the federal government. Not surprisingly, some state governments are supportive of GE crops, while others oppose or remain neutral.

Table 7.2. Institutional framework for regulation of agriculture biotechnology in India. (Deepthi E. Kolady and Shivendra Kumar Srivastava, various sources.)

Category	rDAC	IBSC	RCGM	GEAC	SBCC	DLC
Ethical concern addressed	Biosafety	Biosafety	Biosafety & environmental health	Biosafety & environmental health	Biosafety & environmental health	Biosafety & environmental health
Focus	All emerging recombinant technologies	All recombinant research	All GE organisms in research, use and applications	All GE organisms in research, use and applications	All GE organisms in research, use and applications	All GE organisms in research, use and applications
Role	Advisory at pre-research/use/application stage	Approval for R&D and contained experiments	Monitor ongoing research, develop procedures and protocols, scientific risk assessment, recommend field trials	Recommendation for environmental release, approval for confined field trials, monitor field trials	Monitoring and supervision at post-release stage at state level	Local supervision and compliance
Public comment provision	No	No	No	Yes-recently	No	No
Decision time	Not specified	Not specified	Not specified	Not specified	Not specified	Not specified

In order to take into account these state-level differences, in 2011 the MOEF&CC notified that a 'no objection certificate' (NOC) from respective state governments would be mandatory for the GEAC to issue permits for conducting field trials (MOEF&CC, 2011). This requirement of mandatory NOC potentially triggers 'Federal vs State' conflict based on differences in domestic politics at the state and federal levels (*Business Standard*, 2014) and further complicates the process.

Comparative Analysis of Regulatory Systems and their Responses to Ethical Tensions

In the US regulatory system, each of the three federal agencies has its own independent guidelines and framework to act on specific ethical concerns based on their area of expertise. Additionally, the responsibilities and focus areas are clearly laid out in the Coordinated Framework, which makes the regulatory process more transparent and predictable, with firm timelines for decision-making. Since respective federal agencies follow the guidelines of the Coordinated Framework to take the final regulatory decisions in their areas of authority, the institutional framework in the USA is less susceptible to external political influence. The institutional framework in India differs from that of the USA in many ways. Unlike the more centralized US system, the Indian system has a complicated hierarchical structure with statutory bodies at state and federal levels. Thus, the Indian regulatory system is fragmented functionally and administratively, and provides more opportunities for external influences at various levels than the US system. Additionally, in India, the apex regulatory authority (the GEAC) can only recommend a regulatory action, but the final regulatory decision will be taken by the MOEFF&CC. Compared with the two-party system in the USA, India has a multi-party political system at state and federal levels.

Further, heterogeneity among stakeholders is much higher in India than in the USA. For example, in India technology developers are divided into domestic, foreign, private and public categories; farmers into food and cash crop growers; and consumers into rural and urban categories. The likelihood is very high in India for political parties to exploit the differences in values and interests of each of these stakeholders for their political gains and to exert pressure on the regulator to influence its decision in their favor. Thus, the hierarchical nature of the institutional framework and the role of a political entity in the regulatory decision increase the likelihood of political influence and create further complexity in the Indian regulatory framework and lead to uncertainty in overall regulatory policy. The result is a stagnation in the development and introduction of new GM crops. As evidence, although in 2010 there were 34 ongoing field trials in India, not a single new trait received commercial approval during the last seven years (IGMORIS, 2017).

Ethical concerns and tensions in the USA

Table 7.3 presents some recent examples of ethical concerns raised with respect to the deregulation of GE organisms which contributed to ethical tensions between various stakeholders in the USA. As shown, although there were concerns about regulators' responses to environmental and human health, regulatory agencies followed a science-based permissive approach and enforced the principles of the Coordinated Framework in a predictable way. Additionally, although socio-economic concerns were raised during the review period of GE ethanol corn, because these social justice-related concerns were outside the scope of the Coordinated Framework guidelines, the APHIS deregulated it in 2011. Once deregulated, complaints have to be filed in courts and courts adjudicate on each case. Thus, regulations in the USA follow a risk-based decision-making approach to address general ethical concerns such as environment and human health and food safety, and can be relatively shielded from influences by social justice-related ethical concerns or special ethical concerns.

Table 7.3. Ethical concerns and tensions related to agriculture biotechnology in the USA. (Deepthi E. Kolady and Shivendra Kumar Srivastava, various sources.)

Crop/animal (Ethical concern)	Complaint and action
Bt corn (General concerns on environmental health and biosafety)	• Reports of Bt corn harming monarch butterfly larvae came out in 1999. • USEPA recommended that Bt corn farmers should have buffer zones of conventional corn in their fields.
GE alfalfa (General concerns on environmental health; concerns on regulators' response)	• Complaint in 2007 that APHIS failed to consider environmental effects. • Court injunction against planting or selling until final Environmental Impact Statement (EIS) was prepared. • EIS prepared in 2010. • Court reversed the injunction.
GE sugar beets (General concerns on environmental health; concerns on regulators' response)	• Complaint in 2008 that APHIS violated the National Environmental Policy Act by not conducting an EIS. • EIS prepared in 2012. • APHIS deregulated sugar beets in 2012.
GE ethanol corn (General concerns on environmental health and market; concerns on regulators' response)	• Trade groups and grain processing companies opposed deregulation during review process in 2008. • In 2011, APHIS deregulated it.
GE apples (General concerns on environmental health and food safety)	• In 2010, the company applied for deregulation. • Consumer groups opposed deregulation during public comment period. • FDA deregulated it in 2015.
GE salmon (General concerns on environmental health)	• In 2010, FDA started the regulatory process. • Many, including Congress members and consumer groups, opposed deregulation during review period. • FDA deregulated it in 2015.

However, critics of the current Coordinated Framework raise questions such as whether the current laws are adequate to protect human and environmental health with the new developments in the field, such as plant-based pharmaceuticals, transgenic animals and insects. To ensure that the US regulatory system can adequately assess any risks associated with future GE products, the Obama administration proposed the first comprehensive review of the Coordinated Framework in 30 years in July 2015 (Cowan, 2015). Examples of increasing concern among consumers with respect to the FDA's handling of labeling of GE foods and food products include successful ballot initiatives in states such as Maine and Vermont that require labeling of GE products. To preempt efforts by state and local governments to set their own labeling standards, Congress established a national scheme for the mandatory disclosure of GE ingredients on food labels in 2016. In late July 2016, President Obama signed Senate Bill 764 into law, which requires food manufacturers to disclose the presence of genetically modified ingredients in a text label, a symbol or an electronic or digital link – a QR code – that can be scanned by a mobile device (PBS, 2016).

Ethical concerns and tensions in India

As we show below, although general ethical concerns related to social justice are not the focus of the scientific regulatory process in India, because of the role of a political entity in the regulatory process, these concerns become dominant in the policy discourse and influence the regulatory outcomes. Thus, unlike in the USA, ethical concerns related to social justice lead to ethical tensions among and between various stakeholders in India and influence regulatory outcomes (Table 7.4). Given that there are differences in the institutional framework of regulation, political structure and prevailing ethical concerns in the public discourse between the USA and India, we use the case examples of Bt cotton, Bt eggplant and GE mustard to examine the relationship between ethical concerns and ethical tensions in India and regulators' response to such ethical tensions.

Bt cotton

Cotton is one of the most important cash crops in India and is grown over 13.08 million hectares, primarily by smallholders. Cotton

bollworm complex emerged as one of the major pests of cotton, leading to overuse of insecticides and high crop loss due to insect damage. Although Monsanto approached the DBT in 1990 for transfer of technology and training for Indian personnel to develop GE cotton for protection against bollworm infestation, the Indian government rejected Monsanto's offer, for two reasons: (i) the technology transfer fees demanded by Monsanto were very high (Bhargava, 2003; Ramanna, 2006) and (ii) the strategy of back crossing an American variety to a local variety was rife with problems associated with traditional breeding programs (Bharathan, 2000). To address this, Monsanto built an alliance with the Indian seed company MAHYCO and formed Mahyco-Monsanto Biotech Ltd (MMB). In 1995, the Indian government granted permission to MMB to develop Bt cotton. Bt cotton was expected to

Table 7.4. Ethical concerns and tensions on agriculture biotechnology in India. (Deepthi E. Kolady and Shivendra Kumar Srivastava, various sources.)

Crop (Ethical concern)	Ethical tension/action
Bt cotton (General concerns on biosafety, environmental health and social justice; concerns on regulators' responses)	• R&D started in 1995. • Developer of the technology completed required biosafety studies in 2001. • GEAC did not take a final decision in 2001. • In late 2001, reports of cultivation of illegal Bt cotton came out. • Court ordered destruction of illegal plants and sued the company which produced it. • Regulatory approval in 2002.
Bt eggplant (General concerns on biosafety, environmental health, food safety, shifting power relations, social justice; concerns on regulators' responses)	• R&D started in 2000. • GEAC recommended commercial release in 2009. • Minister of Environment, Forests & Climate Change imposed moratorium on commercial release of Bt eggplant in 2010. • In 2012, the parliamentary panel recommended discontinuation of field trials of all GE crops. • In 2012, the Supreme Court-appointed Technical Expert Committee recommended ban on all field trials and 10-year moratorium on commercialization of GE crops
GE mustard (General concerns on biosafety, environmental health, food safety, shifting power relations, social justice; concerns on regulators' responses)	• Small-scale open field trial started in 2003[a]. • Developer of the technology submitted application for environmental release in 2016. • GEAC recommended commercial release in May 2017.

[a]It suggests that the R&D process started much earlier than 2003, exact year is not available to us.

provide inbuilt resistance against the boll-worm complex, resulting in reduced use of insecticides and increased yields due to the avoidance of crop loss. Even after six years of elaborate tests, MMB was asked to conduct further tests to address ethical concerns related to biosafety.

However, in 2001, the regulators discovered that illegal Bt cotton was grown on more than 11,000 hectares in the state of Gujrat. Citing environmental and biosafety concerns, the GEAC ordered the state government to burn the illegal Bt cotton on fields (Jayaraman, 2001). The GEAC also ordered the state government to keep the already harvested lint under safe custody and conduct animal feeding trials and allergenicity tests before releasing it into the market. However, due to the fear of backlash by cotton farmers, the state government did not follow GEAC orders. The GEAC sued Navbharat Seeds, the developer of illegal seeds, for developing a Bt hybrid without GEAC approval and marketing it without the environmental safety assessment and approval. Although concerns such as Monsanto taking control of the Indian seed industry and influencing farm structure and food sovereignty, and private-sector influencing public research and regulation were prevalent in the public debate on Bt cotton, in the backdrop of the illegal Bt cotton controversy, the GEAC was forced to approve commercialization of Bt cotton in 2002 without any further delay.

Despite the strong empirical evidence for the positive effects of Bt cotton in terms of increased yields, reduced insecticide costs and increased farm income (Kathage and Qaim, 2012; Srivastava and Kolady, 2016), Bt cotton is still at the center of many controversies. Although not verified, there were reports of mass deaths in sheep and goat feeding on Bt cotton from the state of Andhra Pradesh in India and a surge in farmer suicides in states like Andhra Pradesh and Maharashtra (Bharathan, 2000; Naik *et al.*, 2005). Environmental activists and national and international NGOs are using these reports to further exploit the ethical concerns on biosafety and social justice to their favor and to stall the regulatory process in India.

Bt eggplant

Eggplant, locally known as *brinjal*, is an important vegetable crop in India grown mostly by smallholders on 0.68 million hectares. MAHYCO, a domestic seed company, started the R&D of Bt eggplant in 2000. Bt eggplant is expected to provide resistance against eggplant fruit and shoot borer, one of the major pests of eggplant. To address the ethical concerns on food safety and environmental health expressed by NGOs, environmental activists and consumer groups, the GEAC demanded additional test results from MAHYCO, beyond those of the existing guidelines. After examining the biosafety dossier submitted by MAHYCO that included test results from additional studies, the GEAC recommended commercialization of Bt eggplant in 2009.

Consumer groups, domestic and international NGOs, and environmental activists campaigned against the GEAC decision. Unlike cotton, which is a cash crop, eggplant is a food crop. Additionally, the fact that there are other GE food crops in the research pipeline from both the public and private sector raised the stakes of the Bt eggplant case in India. Instead of supporting the GEAC's science-based decision, the Minister of Environment, Forests & Climate Change decided to hold public hearings in seven cities in India. Based on inputs from these consultations, advice from various state governments and comments from national and international scientists, in 2010 the Minister declared a moratorium on the release of Bt eggplant. More specifically, the Minister stated in his decision letter:

> when there is no clear consensus within the scientific community itself, when there is so much opposition from the state governments, when responsible civil society organizations and eminent scientists have raised many serious questions that have not been answered satisfactorily, when the public sentiment is negative and when Bt-brinjal will be the very first genetically-modified vegetable to be introduced anywhere in the world and when there is no over-riding urgency to introduce it here, it is my duty to adopt a cautious, precautionary principle-based

approach and impose a moratorium on the release of Bt-brinjal, till such time independent scientific studies establish, to the satisfaction of both the public and professionals, the safety of the product from the point of view of its long-term impact on human health and environment, including the rich genetic wealth existing in brinjal in our country. (*The Hindu*, 2010, n.p.)

While the GEAC has to follow the 'EPA 1989 Rules' in its decision-making process, no such guidelines are applicable for the Minister. Because of this, it is easy for the Minister to bring in social justice-related concerns into the policy discourse. As shown in Fig. 7.2, the GEAC's science-based recommendation was over-ridden by the Minister, who brought in many social justice-related ethical concerns into the policy discourse. It is also evident from the figure that lobbying by heterogeneous groups of stakeholders influenced the Minister's decision to impose the moratorium.

GE mustard

Indian mustard (*Brassica juncea*) is the third most important vegetable oil source in India. India relies on imports to meet about 50% of its edible oil demand (Sharma, 2014). Mustard is grown on about 6.8 million hectares by smallholders, particularly in rain-fed areas. However, mustard yield is low under rain-fed conditions. Improving vegetable oil production through yield enhancement is an important strategy to reduce reliance on imports and also to improve food security in rain-fed areas. Development of improved varieties and hybrids is one of the approaches to improve mustard yields. Development of hybrids is hard for mustard, a self-pollinated crop. The Center for Genetic Manipulation of Crop Plants (CGMCP) used genetic engineering to develop male sterile and restorer fertility lines to produce hybrid genes. The three transgenes used in GE mustard – *barnase*, *barstand* and *bar* – were developed by Belgian scientists in early 1990s. These same genes are used in GE canola in the USA and Canada (MOEF&CC, 2016). The research in India was financially supported by the DBT and National Dairy Development Board. After more than a dozen years of research, the CGMCP submitted its application for environmental release of GE mustard in January 2016. The GEAC constituted a sub-committee to review the technical details and biosafety dossier submitted by the

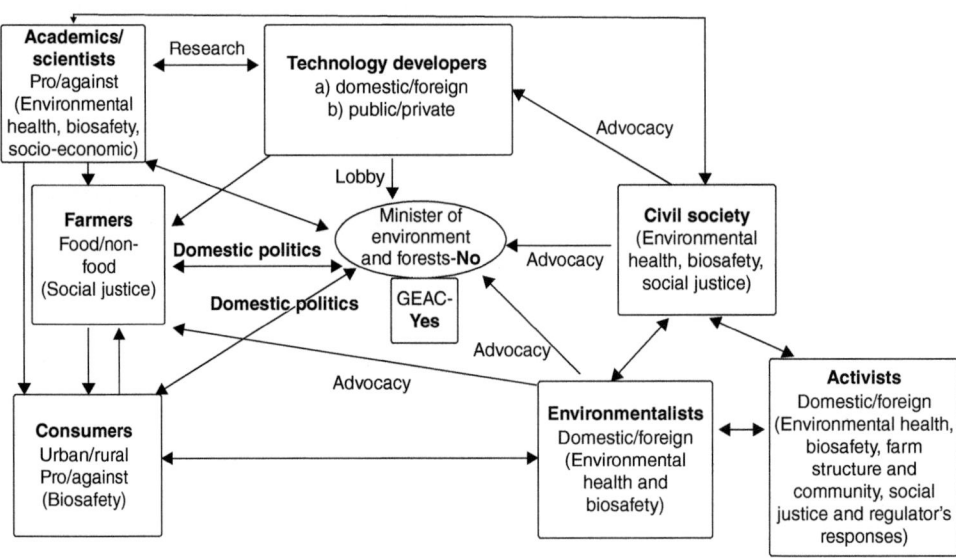

Fig. 7.2. Pathways in which ethical concerns influence the regulatory decisions on Bt eggplant. (Deepthi E. Kolady and Shivendra Kumar Srivastava.)

CGMCP. The sub-committee evaluated the application and prepared a report on Assessment of Food and Environmental Safety (AFES) in September 2016, which concluded that GE mustard is safe for human consumption and poses no risk to the environment. In order to address the earlier concerns on lack of transparency and public consultation, the sub-committee uploaded the report on the Environment Ministry's website for 30 days for wider consultation. The public were allowed to review the full application dossier in person by visiting the MOEF&CC in New Delhi. A total of 759 comments were received from various stakeholders on the AFES report (PIB, 2016). On 12 May 2017, the GEAC recommended GE mustard for commercial release (The Economic Times, 2017). An intense public debate involving scientists, environmental activists, consumer group advocates, politicians and farmers groups followed the GEAC's decision.

The fact that GE mustard is developed by the Indian public sector should have addressed the social justice-related concerns over private-sector and multinational corporations taking control of the Indian seed industry and exploiting smallholders in India. The present government of India has launched the Made in India campaign and the development of GE mustard fits well with that political agenda. However, one of the latest concerns is that the *bar* gene used as a marker imparts resistance to glufosinate, a herbicide marketed by Bayer. Environmental activists argue that the researchers did not reveal this information in their application and use of GE mustard will increase the use of pesticides which will benefit multinational agrochemical companies. Some scientists support this argument. Thus, there is no political or scientific consensus on the biosafety and environmental safety of GE mustard which further delays the regulatory decision on GE mustard.

Conclusions

While the US regulatory system is centralized, with clear guidelines for each of the three federal agencies, the Indian system is fragmented and hierarchical, making it susceptible to domestic politics. Since the federal agencies in the USA can act independently of the legislature and the White House, in general, the regulatory decisions in the USA are shielded from domestic politics. Thus, while there are opposing interest groups who try to exert pressure on regulatory agencies, the US regulatory agencies are able to follow the guidelines in the Coordinated Framework and make science-based decisions in a relatively predictable way. However, in India, the apex regulatory body can only give its recommendations; the final decision will be taken by a political entity at the federal level. We conclude from our analysis that while ethical concerns on environmental and human health and regulators' responses are prevalent in both the USA and India, because of the dominant role of a political entity in India's regulatory process, social justice-related concerns are more prevalent in India than in the USA.

Additionally, our analysis shows that the US regulatory agencies address ethical concerns related to environmental and biosafety issues via risk-based environmental impact assessments and food and feed safety assessments, and recommend risk management strategies in a timely manner. The heterogeneity of stakeholders and the diversity of prevalent ethical concerns coupled with the influence of domestic politics make the regulatory process in India complex and uncertain. As evident from the Indian case examples, there is no silver bullet to address the ethical concerns related to social justice aspects, particularly when stakeholders are very heterogeneous and political stakes are high. Indian regulators' actions in the case of Bt cotton, Bt eggplant and GE mustard highlight the unpredictability of the current regulatory framework. Standard political economic theories, including those of economic interest groups, cannot fully explain the regulatory uncertainty in India. Instead, our analysis shows that types of prevalent ethical concerns, differences in institutional structure and domestic politics contribute to divergence in policy and regulation of agriculture biotechnology in the USA and India.

As technologies evolve and industries mature, public attitudes may shift based on prevailing socio-economic situations, resulting in certain ethical concerns increasing in importance relative to others. It is too early to say whether those shifts in public attitudes will lead to convergence or divergence in policy and regulation of agricultural biotechnology across the globe.

References

Aggarwal, M. (2016) Politics around GM mustard issue heats up. Available at: http://www.livemint.com/Politics/SRbyJxxM119gQf0myFrGmL/Politics-around-GM-mustard-issue-heats-up.html (accessed 1 May 2017).

Berg, P. (2008) Meetings that changed the world: Asilomar 1975: DNA modification secured. *Nature* 455, 290–291. doi:10.1038/455290a

Bernauer, T. and Meins, E. (2003) Technological revolution meets policy and the market: Explaining cross-national differences in agricultural biotechnology regulation. *European Journal of Political Research* 42, 643–683. doi:10.1111/1475-6765.00099

Bharathan, G. (2000) Bt cotton in India: Anatomy of a controversy. *Current Science* 79, 1067–1075. Available at: http://www.jstor.org/stable/24104360 (accessed 5 July 2018).

Bhargava, P. (2003) High stakes in agro research: Resisting the push. *Economic and Political Weekly* 38, 3537–3542. Available at: http://www.jstor.org/stable/4413930 (accessed 5 July 2018).

Business Standard (2014) GM. Field trials: Regulator proposes but most states decline. Available at: http://www.business-standard.com/article/current-affairs/gm-field-trials-regulator-proposes-but-most-states-decline-114032500905_1.html (accessed 1 December 2016).

Chauthaiwale, V. (2016) Cotton, mustard, two GM debates. Available at: http://indianexpress.com/article/opinion/columns/cotton-mustard-two-gm-debates-3004656/ (accessed 1 May 2017).

Cowan, T. (2015) Agricultural biotechnology: Background, regulation, and policy issues. Congressional Research Service Report. Available at: https://fas.org/sgp/crs/misc/RL32809.pdf (accessed 1 December 2016).

Department of Biotechnology (DBT) (2016) Biotechnology: An agent for sustainable socio-economic transformation. Available at: http://www.dbtindia.nic.in/wp-content/uploads/biotechnology-an-agent-for-sustainable-socio-.pdf (accessed 1 December 2016).

FAO/WHO (2015) *Codex Alimentarious Commission Procedural Manual*, 23rd edn. Available at: https://www.fsis.usda.gov/wps/wcm/connect/c9caaaaf-ebc9-437b-b587-e98ca8cc88bf/CACProceduralManual_23e.pdf?MOD=AJPERES (accessed 1 June 2017).

Graff, G.D., Hochman, G. and Zilberman, D. (2009) The political economy of agricultural biotechnology policies. *AgBioForum* 12, 34–46. Available at: http://agbioforum.org/v12n1/v12n1a04-graff.htm (accessed 5 July 2018).

Hochman, G., Rausser, G., Sexton, S. and Zilberman, D. (2008) Agriculture biotechnology in California and the EU. Working paper, Institute of Government Studies, UC Berkeley, California. Available at: http://escholarship.org/uc/item/7kv3s4mg#page-4 (accessed June 2017).

Indian GMO Information System (IGMORIS) (2017) Status of GMOs and products. Available at: http://www.igmoris.nic.in/field_trials.asp (accessed 15 January 2017).

James, C. (2015) *20th Anniversary (1996 to 2015) of the Global Commercialization of Biotech Crops and Biotech Crop Highlights in 2015*. ISAAA Brief No. 51. ISAAA, Ithaca, New York.

Jayaraman, K.S. (2001) Illegal Bt cotton in India haunts regulators. *Nature Biotechnology* 19, 1090. doi:10.1038/nbt1201-1090

Kathage, J. and Qaim, M. (2012) Economic impacts and impact dynamics of Bt (*Bacillus thuringiensis*) cotton in India. *Proceedings of National Academy of Science* 109, 11652–11656. doi:10.1073/pnas.1203647109

Ministry of Environment, Forests & Climate Change (MOEF&CC) (2011) Report to the people on environment and forests 2010–11. Available at: http://envfor.nic.in/sites/default/files/Report-To-The-People-on-Environment-And-Forests%20-2010-11.pdf (accessed 1 October 2016).

Ministry of Environment, Forests & Climate Change (MOEF&CC) (2016) Assessment of food and environmental safety for environmental release of genetically engineered mustard (*Brassica juncea*) hybrid DMH-11 and use of parental events (varunabn 3.6 and EH2 modbs2.99) for development of new generation hybrids. Available at: http://www.moef.gov.in/sites/default/files/Safety%20assessment%20report%20on%20GE%20Mustard_0.pdf (accessed 1 December 2016).

Naik, G., Qaim, M., Subramanian, A. and Zilberman, D. (2005) Bt cotton controversy: Some paradoxes explained. *Economic and Political Weekly* 40, 1514–1517. Available at: http://www.jstor.org/stable/4416465 (accessed 5 July 2018).

National Academies of Sciences, Engineering, and Medicine (NAS) (2016) *Genetically Engineered Crops: Experiences and Prospects*. The National Academies Press, Washington, DC.

Posner, R.A. (1974) Theories of economic regulation. *The Bell Journal of Economics and Management Science* 5, 335–358.

Press Information Bureau (PIB) (2016) GEAC sends public comments on assessment of food and environmental safety report on environmental release of GE mustard to subcommittee. Available at: http://pib.nic.in/newsite/PrintRelease.aspx?relid=151506 (accessed 1 November 2016).

Public Broadcasting System (PBS) (2016) House passes bill to prevent mandatory GMO food labelling. Available at: http://www.pbs.org/newshour/rundown/house-passes-bill-prevent-mandatory-gmo-food-labeling/ (accessed July 2017).

Ramanna, A. (2006) India's policy on genetically modified crops. Working Paper 15. Asia Research Centre, LSE, London. Available at: http://www.lse.ac.uk/asiaResearchCentre/_files/ARCWP15-Ramanna.pdf (accessed 1 November 2017).

Sharma, V.P. (2014) *Problems and prospects of oilseeds production in India*. Centre for Management in Agriculture, Indian Institute of Management, Ahmedabad, India.

Srivastava, S.K. and Kolady, D. (2016) Agricultural biotechnology and crop productivity: Macro-level evidences on contribution of Bt cotton in India. *Current Science* 110, 311–319. Available at: http://www.currentscience.ac.in/Volumes/110/03/0311.pdf (accessed 5 July 2018).

The Economic Times (2017) GEAC clears GM mustard for commercial use. *The Economic Times*, 12 May. Available at: https://economictimes.indiatimes.com/news/economy/agriculture/geac-clears-gm-mustard-for-commercial-use/articleshow/58634821.cms (accessed 1 September 2017).

The Hindu (2010) Bt Brinjal: Note by Ministry of Environment and Forests. Available at: http://www.thehindu.com/news/national/Bt-Brinjal-Note-by-Ministry-of-Environment-and-Forests/article16578296.ece (accessed 1 October 2017).

Thompson, P. (2000) Food and agricultural biotechnology: Incorporating ethical considerations. Available at: http://www.iatp.org/files/Food_and_Agricultural_Biotechnology_Incorporat.htm (accessed December 2016).

Zusman, P. (1976) The incorporation and measurement of social power in economic models. *International Economic Review* 17, 447–462. Available at: http://www.jstor.org/stable/2525712 (accessed 5 July 2018).

8 Technological Pragmatism: Navigating the Ethical Tensions Created by Agricultural Biotechnology

Dane Scott*

Mansfield Center, University of Montana, Missoula, Montana, USA

Introduction

Many critics of genetically engineered (GE) crops and livestock often label them 'techno-fixes'. 'Technological fix' criticisms brand innovations as superficial solutions that do not get at the root of problems but rather create new ones. The rotten roots of problems, which technological fixes fail to address, are identified as social and political in nature. The typical use of a 'technological fix' criticism against GE crops is illustrated in Greenpeace International's campaign against Golden Rice. Golden Rice is a biofortified strain of GE rice designed to address vitamin-A deficiency (VAD), which is widespread among poor populations who get most of their calories from rice. Greenpeace.org's special report on Golden Rice states: 'GE "Golden" rice does not address the primary causes of VAD, which are poverty and lack of access to a healthy and varied diet. Thousands of children die or go blind each year because of Vitamin A deficiency diseases (VADD). GE rice is a *technological fix* that may generate new problems' (Greenpeace, 2013, n.p.; emphasis added). Greenpeace's controversial campaign against Golden Rice asserts that the proposed applications of this technological fix are morally wrong. In their view, VAD is a social and political problem, not a technological one: the root causes

of VAD are poverty, unjust social and economic structures, and this technological fix risks dangerous health and environmental side effects. Nonetheless, many of modern civilization's most powerful institutions – life science corporations and pharmaceutical companies, research institutes and universities – are designed to generate technologies to tackle a wide array of social problems in the arenas of medicine, agriculture, energy, and so forth. On the one hand, supporters of agricultural biotechnology believe this frontier of science and technology will usher in a new era of human progress – that is, a biotech revolution. On the other, detractors like Greenpeace see GE crops and animals as generating another round of flawed technological fixes that will accelerate the vicious cycle of health and environmental problems resulting from industrial agriculture. These opposing views have contributed to ethical tensions over innovations in agricultural biotechnology. The goal of this inquiry into 'technological fix' criticisms is to identify a philosophy of technology with the conceptual tools to reduce the ethical tensions created by specific innovations in agricultural biotechnology.

An important source of ethical tensions over agricultural biotechnology is a polarizing conflict between two opposing philosophies of technology, which can be labeled

* E-mail: dane.scott@mso.umt.edu

technological optimism and technological pessimism (Crabill *et al.*, 2012). Technological optimists exhibit high levels of trust in new biotechnologies: they are committed to the idea that technological innovations are essential to building a better world. From this view, the biotech revolution in agriculture represents progress toward conquering humanity's historical enemies of pest, pestilence and famine. Technological pessimists, meanwhile, exhibit strong attitudes of distrust. They are committed to the idea that technological civilization is on a dangerous course by dominating nature with its powerful innovations. From this view, the biotech revolution represents a serious threat to the future of humanity and the earth's biodiversity. Those who hold strong optimistic or pessimistic philosophies of technology do not experience serious ethical tensions over agricultural biotechnology; their respective philosophies allow them to see this issue with the moral clarity of black and white. However, most people find themselves philosophically in the muddled middle on the GE controversy – that is, somewhere along the spectrum between the poles of techno-optimism and techno-pessimism. Those of us who reside in there must wrestle with the ethically ambiguous nature of new agricultural biotechnologies, which come with benefits as well as harms, goods as well as evils. Thus, another ethical tension is that while most people appear to fall somewhere in the middle of the debate, the debate itself occurs on the extremes.

This chapter draws from and builds on Scott (2011) by providing a path through the thicket of ethical tensions created by innovations in agricultural biotechnology. In the first section I offer a better understanding of the notion of a technological fix and technological fix criticisms. Then I will clarify and evaluate technological optimism and technological pessimism, respectively. Following that discussion I will argue for a technological pragmatism that provides for a more discriminating approach to the tensions created by innovations in agricultural biotechnology. In the final section I will illustrate how the conceptual tools provided by technological pragmatism's critique of

technological fixes might be used to guide more ethically conscious research into GE crops and animals.

Technological Fix Criticisms and Wicked Problems

The idea behind 'technological fix' criticisms is that certain problems are essentially social and political, not technological. However, the notion that there are definitive causes to the big problems humanity faces in the 21st century is dubious. Problems like chronic poverty, world hunger and climate change have been labeled 'wicked problems' (Rittel and Webber, 1973). Wicked problems are ones where 'important values are at stake, factual issues are shrouded in uncertainty, options for moving forward are mutually exclusive and have irreversible consequences, but there is no fundamental agreement on what the problem is' (Thompson, 2014, p. 7). The sources of wicked problems defy definitive identification and the solutions are controversial. When, where and how we should address wicked problems involves a complex mix of ethical, political, economic, practical, technological and scientific considerations that are clouded in a haze of ambiguity.

Given that the complex and indefinite character of many of the 21st century's big problems, such as poverty, malnutrition, pollution, climate change, massive aquatic dead zones and high rates of loss of biodiversity, are in various ways connected to agriculture, in specific situations a technological fix may be the best that can be done. There are numerous instances where technological fixes to difficult problems with crucial social and political factors are seen as the best available option. For example, the global climate crisis is very much driven by social and political factors. One could say that the threats of dangerous climate change are caused by political failures to implement timely and effective policies to mitigate greenhouse gases. Because of this political failure, many coastal cities are confronted with rising sea levels and destructive storm surges. These cities may decide that their best available option is to adapt by construct- ing floodgates and sea walls, which can be characterized as technological fixes because they are addressing the 'symptom' of rising sea levels and not the 'disease' of entrenched social, political and economic commitments to greenhouse gas-producing energy sources. In situations where problems are severe and imminent, and social and political solutions are slow and uncertain, expedient technolog- ical fixes may be the best available option. Nonetheless, there is a danger in modern technological civilization of overusing tech- nological fixes at the expense of social and political solutions. This is particularly the case in modern agriculture where institu- tional inertia in research and business and entrenched habits of thought automatically approach problems as technological puzzles with technological solutions. There is an im- plicit commitment to technological optimism and a bias toward technological fixes built into modern civilization that should be identified and challenged.

While the term 'techno-fix' has become a derisive label, it started as a recommenda- tion for a positive course of action. Alvin Weinberg (1969) coined the term in his book *Reflections on Big Science*. Weinberg argued that while technological fixes cannot replace social engineering, they should be used as a positive social action. He characterizes a technological fix as the solution to a prob- lem that results from reframing a social and political problem as a technological one. The notion of the technological fix is in reality a practical strategy for addressing intractable social and political problems. The strategy is to reframe seemingly insurmountably com- plex social and political issues as engineer- ing problems. Weinberg lists four benefits of the technological fix strategy. First, technolog- ical problems are much simpler than social problems; they are easily defined and pre- scribed a solution. He writes that 'the availa- bility of a crisp and beautiful technological solution often helps focus on the problem to which the new technology is the solution' (Weinberg, 2001, p. 109). Second, techno- logical problems do not have to deal with the complexity and unpredictability of human behavior. The technological fix strategy, for better or worse, factors out the messiness of politics. Third, they provide policy makers with more options – additional means for addressing social problems. Finally, they can buy time until the problem can be dealt with on a deeper level. In sum, the technological fix idea denotes a problem-solving strategy that reduces the complexity of social and political problems. Once this is done, the only factors that will be considered are those that can be interpreted in terms of a techno- logical system.

The overall impression of Weinberg's writings is that political solutions rarely work, while techno-fixes are quick, efficient and effective. Throughout his career, Wein- berg remained a champion of technological fixes – he subtitled his 1994 memoir, *The Life and Times of a Technological Fixer*. Even so, Weinberg was a realist. He was aware that technological fixes are not ultimate solu- tions and can generate unintended conse- quences. During his distinguished career as a nuclear physicist and the director of the Oak Ridge Nuclear Laboratory (from 1955 to 1973), Weinberg was a persistent advo- cate for nuclear energy. However, he also popularized the 'Faustian bargain' metaphor for nuclear power. In 1972, at the high point of enthusiasm over nuclear energy, Wein- berg writes: 'We nuclear people have made a Faustian bargain with society. On the one hand, we offer ... an inexhaustible source of energy .... But the price that we demand of society for this magical energy source is both a vigilance and a longevity of our social institutions that we are quite unaccustomed to' (1972, p. 33). Weinberg was a firm believ- er in 'big science' (another term he coined) and technological progress, yet he under- stood that powerful modern technologies come with risks and responsibilities that threaten to exceed our grasp.

Technological Optimism, Technological Fixes and Agricultural Biotechnology

Weinberg's techno-optimism is the product of a long philosophical tradition sometimes labeled the Enlightenment Project. This tradition interprets the history of humanity in terms of the philosophical idea of progress. In this view, the story of humanity follows a general trend of progress in conquering the perennial enemies of pest, pestilence, famine, poverty and war through the advance of reason, democracy, science and technology. This interpretation of history is implicitly evident in many books that advocate for agricultural biotechnology. Two examples of this are G. Pence (2002) *Designer Foods* and R.P. Thompson (2011) *Agro-Technology: A Philosophical Introduction*. For Pence and R.P. Thompson (R.P. Thompson is distinguished from Paul B. Thompson, whose work will be discussed later), the purpose of technological agriculture is to increase production and maximize human welfare. Their ethical arguments for agricultural biotechnology are grounded in a progressive interpretation of the history of technology. For example, R.P. Thompson begins his book by quoting Jeffrey Sachs, who writes: 'I believe that the single most important reason why prosperity spread, and why it continues to spread, is the transmission of technologies and the ideas underlying them (2005, pp. 41–42). This optimistic view of the history of technology leads Pence and Thompson to make sweeping arguments for biotechnology as the next step in the history of human progress.

It is a common practice for techno-optimists debating GE crops to cite the Green Revolution as evidence to support their advocacy for agricultural biotechnology. Both Pence and Thompson devote significant portions of their books to celebrating the Green Revolution to justify the 'Gene Revolution'. The Green Revolution denotes the period from the 1960s to the 1980s when technological inputs and high-yield varieties of wheat, rice and corn greatly increased food production in South Asia and South America. The father of the Green Revolution, Norman Borlaug, won the Nobel Peace Prize in 1970 for his efforts in leading this agricultural revolution. In his later years, Borlaug became an outspoken advocate for biotechnology. In an essay with the provocative title, 'Ending world hunger: The promise of biotechnology and the threat of antiscience zealotry', Borlaug (2000) affirms the idea of progress and a commitment to technological optimism as a philosophy of technology. He writes: 'Genetic modification of crops ... is the progressive harnessing of the forces of nature to the benefit of feeding the human race.... The genetic engineering of plants at the molecular level is just another step in humankind's deepening scientific journey into living genomes' (2000, p. 489). In this philosophy of technology, the past successes of scientific and technological innovations justify trust in new technological innovations like biotechnology. An optimistic interpretation of the Green Revolution justifies trust in the next technological revolution in agriculture, the Gene Revolution.

There are, however, more pessimistic interpretations of the history of modern technological agriculture. It is well documented that Green Revolution technologies and practices – high-yield varieties, industrial fertilizers, synthetic pesticides, mechanization, irrigation, monocultures, and so forth – have come with very high environmental costs. These costs include the loss of biodiversity, nitrogen and phosphorus pollution of surface and ground waters leading to massive aquatic dead zones, degradation of soils from overtillage, pollution from chemical pesticides. In a *Nature* article assessing the legacy of intensive production practices, Tilman *et al.* highlight 'the need for more sustainable agricultural methods' (2002, p. 672). Techno-optimists are, of course, aware of these problems but trust that the gene (or biotech) revolution will solve the dangerous side effects of the Green Revolution. For example, in an article arguing that agricultural biotechnology is needed to solve the problem of feeding people and conserving the natural environment, the biotechnologist Anthony Trewavas writes:

All technologies have problems because perfection is not in the human condition. The answer is to improve technology once difficulties appear; not, as some would wish, discard technology altogether. Remove the problems but retain the benefits! The benefits of modern agricultural technology are well understood; now is the time to reduce the undoubted side effects from pesticides, soil erosion, nitrogen waste, and salination. GE technology certainly offers some good solutions. (2001, p. 178)

Techno-optimists trust that progress in biotechnology will correct many of the problems created by Green Revolution technologies. This optimistic philosophy of technology engenders strong attitudes of trust in innovations in agricultural biotechnology. The interpretative schema provided by the idea of progress allows philosophers and scientists like Pence, R.P. Thompson, Borlaug and Trewavas to confidently assert that the humanitarian successes of the Green Revolution justify the Gene Revolution.

Opponents of agricultural biotechnology are quick to use the standard criticisms of techno-fixes to counter enthusiasm for agricultural biotechnology. They assert that the root causes of world hunger, for example, are social and political, *not* technological. Thus, in the long run, GE techno-fixes will make things worse. In an article opposing GE crops as a solution to world hunger, the researcher and alternative agriculture activist Peter Rosset writes, 'The real causes of hunger are poverty, inequality and the lack of access. Too many are too poor to buy the food that is available' (2002, p. 82). According to the United Nation's Food and Agriculture Organization, world agriculture produces enough food to provide every person on the planet with enough calories to meet daily requirements (FAO, 2002). The political problems of chronic poverty and lack of access to arable land are the often-cited reasons for hunger and food insecurity, and not the lack of new technologies. Techno-optimists respond to this standard 'techno-fix' criticism with a standard argument for technological fixes. While strictly speaking there might be enough food to feed the current population, the excess food is in wealthy countries and severe poverty and malnutrition are in poor countries. Further, there are many reasons for chronic poverty and malnutrition involving issues of war and peace, the history of colonialism, corrupt governance, and much more. Solving these multifaceted and intractable political problems is an overwhelming task

with long-term horizons. Solving the technological puzzle of conjuring 'more food from the plants we grow' (Trewavas, 2002, p. 150) can be accomplished in timeframes that matter for millions of hungry people. This type of back and forth between techno-pessimists and techno-optimists over technological fixes is common, and it is indicative of the polarized debate over biotechnology. Further, the contentiousness of the GE debate makes it extremely difficult to assess how biotechnology might be best used to help tame the wicked problems of widespread malnutrition, environmental degradation or global climate change.

By way of summary, techno-optimists interpret the history of modern agriculture in light of the philosophical idea of progress. They experience little ethical tension over agricultural biotechnology because its innovations represent progress toward conquering hunger and malnutrition. But a commitment to the philosophical idea of progress and its implicitly utopian philosophy of technology is based on trust. This trust is supported by a selective interpretation of the history of agricultural technology that highlights the goods and downplays the evils. While an optimistic interpretation of technological agriculture is to a certain degree plausible, rival pessimistic interpretations, which highlight the evils and downplay the goods, counter it. Taken together, the tensions created by rival optimistic and pessimistic interpretations of modern agriculture create an ambiguous and conflicting story that engender ethical tensions over agricultural biotechnology.

Technological Pessimism, Technological Fixes and Agricultural Biotechnology

The philosopher of technology, Langdon Winner, points out the weak supports of techno-optimism's historiography. He writes, 'In the twentieth century it is usually taken for granted that the only reliable sources for improving the human condition stem from new machines, techniques and chemicals. Even the recurring environmental and social ills have rarely dented this faith' (2004, p. 104). In his many writings on the history and philosophy of technology, the M.I.T. historian Leo Marx sought to undermine faith in the idea of progress by pointing to its illusory foundations in 17th-century philosophies of history. Marx provides a sweeping critique of technological fixes as an essential element as part of a discredited conception of history:

> To dismiss the possibility of a scientific or technological 'fix' is a commonplace of contemporary intellectual discourse. But too often the idea is treated as if it were a single, discrete, isolable, vulgar error – a tiny speck of bad thinking easily removed from the public eye. Unfortunately, the dangerous idea of a technical fix is embedded deeply in what was, and probably is, our culture's dominant conception of history. (1983, p. 7)

For Marx, technological fixes are defective because they derive from a discredited philosophy of history committed to a necessary relationship between scientific and technological progress and human progress. Marx writes that 'the assumption is that the achievements of scientists and engineers translate more or less naturally and predictably – in the ordinary course of events – into solutions of such grave problems' (pp. 6–7). He calls this assumption a 'logical abyss in our thinking' and responds to it by asserting that 'few arguments could be more useful today than one aimed at persuading the world that science and technology, essential as they are, cannot save us . . . . the most urgent problems on the human agenda inhere in the man-made, not the natural environment. They are political, not scientific, and thus scientific progress cannot be the basis for their resolution' (p. 8). For Marx, the philosophical criticisms of technological fixes point to the Enlightenment's discredited philosophy of history that is at the foundation of the technological optimists' worldview, which creates a strong bias to trust innovations in agricultural biotechnology. According to Marx, the standard criticisms of techno-fixes hold true: the big problems humanity now faces are not technological but social, political and moral.

Another cultural historian, Lynn White, provided an influential moral critique of the idea of progress and technological fixes as the 'domination of nature'. White's (1967) much cited and discussed essay, 'The historical roots of our ecological crisis', appeared at the right moment for maximum influence, just as the environmental movement was entering popular consciousness in the 1960s. His essay developed several themes that contributed to techno-pessimism as a philosophy of technology. The most important theme is the moral condemnation of an anthropocentric worldview that justifies the domination of nature. White's thesis is that the origins of the 20th century's ecological crisis are found in Western Christianity's anthropocentric conception of the relationship of humans to nature that is merged with powerful scientific and technological developments. He diagnosed the 'root' cause of the mid-20th century environmental crises as the merging of the power created by science and technology with a dominant biblical worldview that justifies the use of that power to dominate and control nature. White reinterprets the history of technological agriculture in dark, pessimistic terms by applying the sweeping moral critique of idea of progress and its technological fixes as the domination of nature. White speculates that the development of the heavy plow in northern Europe was not a leap forward in human progress to satisfy hunger, but a significant development in the alternative story of human domination of nature. The heavy plow, with its vertical knife blade, sliced deep into the rich, wet soils of northern Europe, opening up the earth. In White's pessimistic interpretation of the history of agriculture, prior to the heavy plow, 'man had been part of nature; now he was the exploiter of nature' (1967, p. 1205). Like numerous environmental critics of technological civilization, White references an important building block in the philosophy of techno-optimism, Francis Bacon's 17th-century utopian vision in *The New Atlantis*. The creed of Bacon's utopia is 'scientific knowledge means technological power over nature' (White, 1967, p. 1203). White's moral critique of the ideas of progress and techno-fixes aims to undermine the views that technological power is essentially a benign and positive force. The history of agricultural technology, in White's interpretation, is a story of how humans developed an immoral and dysfunctional understanding of the human–nature relationship in terms of domination.

There are many environmental writers and thinkers in this tradition who provide pessimistic interpretations of the history of agriculture to rival the optimistic interpretations. Very briefly, two examples are Paul Shepard and Dave Foreman. In his book, *Coming Home to the Pleistocene*, Shepard asserts, 'If there is a single complex of events responsible for the deterioration of human health and ecology, agricultural civilization is it' (1998, p. 103). In his autobiography, Dave Foreman writes that with the advent of agriculture came an ever-widening rift between the 'wilderness that created us and the civilization created by us' (1991, p. 69). In this view, the birth of agriculture and the domestication of plants and animals did not represent human progress. Rather, the farms are seen as the battlefields for a war against nature.

The moral critiques of 'anthropocentrism' and the 'domination of nature' are key features in a technological pessimist's interpretation of agricultural biotechnology. For example, the philosopher Keekok Lee (1999) uses the idea of the domination of nature to interpret the moral significance of agricultural biotechnology. Like many environmental philosophers, Lee builds her arguments against biotechnology by employing the Aristotelian distinction between natural objects and artifacts. An artifact would be any object that has been manipulated for human purposes (e.g. a piece of obsidian rock that is shaped by human hands for the purpose of hunting), while a natural object would be formed by unhindered natural forces (e.g. a piece of obsidian that is shaped by geological forces only). Lee sees agricultural biotechnology as an ontological threat to nature because it turns natural genomes into human artifacts. In terms of the domination of nature, Lee argues that GE crops engineered for pesticide traits are more ethically pernicious than synthetic pesticides. In her view,

synthetic, chemical pesticides are 'nature polluting'; they are harmful to the environment but their effects can be mitigated and reversed. However, the philosopher, Christopher Preston, points out that in this view genetic engineering 'manipulates nature at such a fundamental level [that it poses] an altogether different kind of problem' (2008, p. 28). By manipulating nature at the level of the DNA molecule, humans are 'systematically transforming naturally occurring beings ... to become artificial ones' (Lee, 1999, p. 1). Preston remarks that Lee seems to think that, '*Bt* corn is ... the deepest kind of biotic artifact. Its very genome is the product of human intention' (2008, p. 28). Nature-replacing biotechnologies represent a threat to the ontological category of natural value. Biotechnology systematically and irreversibly replaces morally valuable natural objects with less valuable artifacts. In this pessimistic view, GE crops are interpreted as artificial creations of the morally misguided anthropocentric worldview and the domination of nature.

By way of summary, techno-pessimism, as a philosophy of technology, interprets the history of modern agriculture in light of the moral critique of anthropocentrism and the domination of nature. From this view there can be little ethical tension over agricultural biotechnology because its innovations are instruments of a morally misguided worldview. This philosophy of technology engenders strong attitudes of mistrust of agricultural biotechnology. In as much as technological pessimism offers a positive philosophy, it is in the form of an Arcadian myth of returning to nature through simple living and organic farming, on an earth with a greatly reduced human population. However, it is an error to replace the philosophical conceit of technological utopianism with the philosophical conceit of back-to-nature Arcadianism. Neither philosophy can serve as a guiding vision for taming the wicked problems of feeding ten billion people and successfully navigating numerous global environmental challenges such as climate change, massive aquatic dead zones and high rates of loss of biodiversity, to name a few. For the vast majority, the 'just say no' attitude to agricultural biotechnology of techno-pessimism, like the 'just say yes' of techno-optimism, fails to offer conceptual tools needed to resolve the ethical tensions created by agricultural biotechnology. A more pragmatic philosophy of technology is needed.

Technological Pragmatism, Technological Fixes and Agricultural Biotechnology

Leo Marx's critique of technological fixes focuses on undermining faith in the philosophical idea of progress, which still holds sway over technological civilization. Marx argues that in the face of contemporary social and environmental problems, it is no longer tenable to believe that scientific and technological progress naturally leads to human progress. However, there are pragmatic conceptions of scientific progress that do not depend on the Enlightenment Project's philosophical idea of progress. In the late 20th century, historians and philosophers of science began to reevaluate the idea of scientific advancement. By far the most influential book of this period was Thomas Kuhn's (1996) *The Structure of Scientific Revolutions*. Kuhn challenged the cumulative thesis, which holds science is coming increasingly closer to understanding the ultimate structure of reality, with a revolutionary account, which holds old theories are overturned and replaced with new ones. There are numerous interpretations of Kuhn's work and his thesis remains controversial, particularly regarding the imprecise notion of a paradigm. However, it is safe to say that, after Kuhn, it is no longer possible to hold an uncritical belief in scientific progress as the cumulative advance of the scientific enterprise. Rather, it is circumscribed within a paradigm that helps create a worldview and culture. Scientific progress is best characterized pragmatically in terms of the successes of normal science as a puzzle-solving activity, which is guided by a research paradigm or, as Kuhn later calls it, a disciplinary matrix. Since modern technology is intrinsically linked to

modern science, challenges to the philosophical idea of scientific progress have consequences for techno-optimism. If science is not necessarily progressive, neither are the technologies that are based on its theories.

In his pioneering book, *Spirit of the Soil*, Paul B. Thompson (distinguished from the R.P. Thompson mentioned earlier) employs Kuhn's notion of a paradigm to develop a philosophical critique of modern agricultural research. Thompson's philosophy of techno-pragmatism (if you will) provides a much more nuanced and useful critique of technological fixes and agricultural biotechnology than the polarizing, 'just say yes' of techno-optimism and 'just say no' of techno-pessimism. Technological pragmatism is capable of providing the conceptual tools to navigate the thicket of ethical tensions created by agricultural biotechnology. In *Spirit of the Soil*, Thompson criticizes the dominant research paradigm and institutional culture that narrowly focus on developing technologies aimed at increasing production. In a chapter titled 'The Productionist Paradigm', Thompson describes the worldview and culture created by that paradigm:

> Agricultural scientists regard their work as successful when it is widely adopted, and the surest path toward adoption is to increase productivity of farming operations.... Agricultural disciplines, departments, and universities are measured by their success in the creation of production enhancing technology. (1995, p. 68)

The cumulative effect of productionism 'is an industrial agriculture for which the goal of making two blades grow where one grew before is never questioned, where those who succeed at this quest are bestowed with honors, and where those who fail to take it up are regarded with puzzlement' (p. 67). The productionist worldview governs research and development that is characterized by a 'total confidence in production enhancing agricultural technologies' (p. 60). In his description and critical analysis, Thompson identifies the ethical and philosophical features of the productionist paradigm and demonstrates that they cannot hold up to philosophical scrutiny. Nevertheless, productionism continues to barrel ahead, generating technological fixes because of institutional inertia and entrenched habits of thought.

The productionism generates techno-fixes, and is open to the criticisms of this strategy. As Thompson notes, this research paradigm follows the pattern of 'aggressive applied scientific research, followed by equally aggressive effort of technological transfer ... [that has become] the headlong and unreflective application of industrial technology for increasing production' (1995, pp. 45, 91). Within the productionist paradigm, the problems of agriculture are framed as engineering puzzles with technological solutions. The major defects with this approach arise from its narrowness. By too narrowly defining problems in terms of production and solutions in terms of technology, this paradigm excludes important social, political, ecological and evolutionary factors that contribute to these problems and ignores alternative solutions. Further, because this strategy fails to account for important contributing factors and to consider alternative solutions, it frequently generates harmful, unintended social and environmental consequences.

Thompson's philosophical pragmatism criticizes productionism and the technological fix strategy while also rejecting the sweeping, ideological criticisms of techno-pessimism. More specifically, his critical analysis of the productionist paradigm identifies its defects and offers avenues for correcting these defects in future research into GE crops and livestock. He also critiques technological pessimism and its rejection of agricultural biotechnology. Thompson identifies two implications of the ecocentric ethic of the pessimistic view that sees 'agriculture as inimical to nature preservation' (2003, p. 192). Techno-pessimism, as described previously, is 'committed to the twin goals of (1) minimizing the extent of land given over to agriculture, and (2) isolating agriculture from the rest of nature'. Thompson observes that when followed to its logical conclusion, 'this is the philosophical recipe for the most thoroughly industrialized agriculture we can possibly imagine' (p. 192). Further, arguments against agricultural biotechnology using the natural/artifactual distinction are confused and

bewildering. It is unclear how someone using the natural/artifactual distinction could accept conventional plant breeding, without which agriculture and civilization would not exist, and reject biotechnology. If one rigorously applies the natural/artifactual distinction, all agricultural crops and domesticated animals, not just the products of genetic engineering, bear the mark of human purpose through thousands of years of selective breeding. This line of argument, which is consistent with White's and Shepard's ecocentric ethics, makes all agricultural crops and domesticated animals an ontological threat against nature. The logical, and fantastical, conclusions of this line of argument would be to abandon agriculture, abolish civilization and return to hunting and gathering – back to the Pleistocene.

As was seen earlier, the Green Revolution illustrates how productionism and the technological fix strategy dominate modern agriculture. This important episode in the history of agriculture has been lauded by optimists and condemned by pessimists. Both assessments cannot be right and the truth is somewhere in between. In many ways, the Green Revolution was an enormous success. But its technologies came with trade-offs. The high-yield varieties, inorganic fertilizers, synthetic pesticides solved the problems for which they were designed, increasing production. However, they did not solve the social and political problems that contribute to hunger and malnutrition because the technological fix strategy ignores these as complicating factors. The strength of this strategy in solving complex problems is simplification; it eliminates complicating social and political factors by reframing the problem as an engineering puzzle. The Green Revolution demonstrates that this strength is also a weakness. When one takes a longer and broader view, techno-fixes tend to delay and relocate problems. The Green Revolution did not solve the problem of hunger and malnutrition in South Asia; it delayed it and transformed it. For example, in his last years, Norman Borlaug argued that wide application of agricultural biotechnology is necessary to increase production to keep pace with population growth. Borlaug writes that, 'Despite the success of the Green Revolution, the battle to ensure food security for hundreds of millions of miserably poor people is far from won. Mushrooming populations, changing demographics, and inadequate poverty intervention programs have eroded many of the gains of the Green Revolution' (2000, p. 487). In addition to its inability to keep pace with population growth, Green Revolution technologies relocated and transformed problems by creating new ones. Again, it is true that these technologies raised yields and helped feed millions of people who might have otherwise died of hunger and malnutrition; it is also true that they generated many negative social and environmental unintended consequences.

In fact, taking a longer and wider view, the negative environmental consequences of modern technological agriculture threaten to undermine its important successes in raising yields and lowering food costs of billions of people. In a comprehensive review of the environmental consequences of the Green Revolution, Evenson and Gollen write that the critics of the Green Revolution have 'raised concerns about the sustainability of intensive cultivation – e.g., the environmental consequences of soil degradation, chemical pollution, aquifer depletion, and soil salinity – and about the differential socioeconomic impacts of new technologies. These are valid criticisms' (2003, p. 761). David Tilman observes that 'it is unclear whether high-intensity agriculture can be sustained, because of the loss of soil fertility, the erosion of soil, the increased incidence of crop and livestock diseases, and the high energy and chemical inputs associated with it' (1998, p. 211). Tilman concludes that 'it is not clear which is greater – the success of modern high intensity agriculture, or its shortcomings' (p. 211). The benefits of raising yields with Green Revolution technologies are the high costs of a legacy of global environmental problems.

Critics of the Green Revolution also criticize agricultural techno-fixes for generating unjust social consequences. In 2004, the United Nations Food and Agriculture Organization (FAO) issued a report titled, *Agricultural Biotechnology: Meeting the Needs of the Poor?* The report held that agricultural biotechnology could benefit poor farmers in developing countries. In response, an open letter to the director of the FAO, signed by 670 nongovernmental organizations, challenged the use of technological fixes to solve problems of hunger and malnutrition in poor countries (see Paarlberg, 2005). The letter makes explicit reference to the Green Revolution as an unsuccessful technological fix that generated negative social side effects.

> [The FAO's report on biotechnology] proposes a technological 'fix' of crops critical to the food security of marginalized peoples … If we have learned anything from the failures of the Green Revolution, it is that technological 'advances' in crop genetics for seeds that respond to external inputs go hand in hand with increased socio-economic polarization, rural and urban impoverishment, and greater food insecurity. The tragedy of the Green Revolution lies precisely in its narrow technological focus that ignored the far more important social and structural underpinning of hunger. The technology strengthened the very structures that enforce hunger. (Organic Consumers Association, 2004, n.p.; emphasis added)

This criticism assumes an essential connection between agricultural biotechnology, the *narrow* technological fix strategy and the productionist paradigm. However, there is no necessary reason for this connection. GE crops and livestock can be designed using other strategies and paradigms. That said, the letter is justified in identifying the narrowness inherent in the techno-fix strategy and the productionism that contributes to social injustices. Paul B. Thompson notes that 'Skeptics of mainstream development and mainstream agricultural science have powerful reasons to believe that it is time for an alternative approach' (2009, n.p.). Alternative approaches should focus on correcting the defects of the technological fix strategy by including social and ecological factors that contribute to problems and including contributing to more just social arraignments as a goal.

One can draw several lessons from a pragmatic interpretation of the Green Revolution that is capable of seeing its successes and failures, benefits and costs. These lessons can be instructive in changing the ways technologies are used to tame the wicked problems of the 21st century. One key lesson is that determining the success of a technological fix depends on who defines success and how it is defined. This explains why modern technological agriculture is so controversial. If success is narrowly defined in terms of increasing production, then this approach can be judged a success. However, if success is defined more broadly to include long-term social and environmental consequences, judgments of success are controversial. The practical criticisms of technological fixes serve as a warning against the inherent dangers of addressing complex, multifaceted problems with the narrow technological fix strategy. This approach only 'solves' problems in terms of the narrow frame of an engineering puzzle. That is, techno-fixes can solve the problem for which they were designed, but when one pans-back to the larger complex, dynamic webs that compose the world of humans and nature, a different judgment might be made. When one takes a wider and longer view, techno-fixes tend to transform, relocate or delay the problems, or create new ones. Scientists, technologists, agronomists and decision makers need to move beyond productionism and characterize their success in much broader terms that include social justice and ecological health, or integrity. Even so, these criticisms do not necessarily require abandoning the technological fix. In certain cases, technological fixes can serve an ameliorative role that may be good enough or all that can feasibly be done. Technological fixes involve trade-offs and side effects, and possibly could result in problems that are worse than the original problem – what Edward Tenner (1997) calls 'revenge effects'.

By concentrating attention on GE crops, techno-optimists and techno-pessimists are focusing on the wrong problem and driving an unproductive, polarized debate. Agricultural biotechnology may provide important possibilities for confronting the serious global problems of this new century, which, again, are nearly all intertwined with agriculture. However, key opportunities will be lost and new social and environmental problems will be created unless an alternative paradigm with alternative strategies is found to replace productionism and the technological fix strategy, whose defects are well documented and understood. With these points in mind, the final section will briefly point out examples where technological pragmatism critique of technological fixes can be used to reduce ethical tensions.

Agricultural Biotechnology, Ethical Tensions and Technological Pragmatism

Agricultural biotechnology has created at least two sources of ethical tensions: fears that GE crops will have unintended social as well as environmental consequences. The negative social impacts are often characterized in terms of social justice, where it is feared that new biotechnologies will drive greater income disparities between rich and poor farmers. As was noted above, these fears are real and well documented. While there is no space to go into detail here, this phenomenon has been labeled the technological treadmill (Cochrane 1958; Levins and Cochrane, 1996) and has been discussed extensively. Productionist, technological fixes tend to favor large, wealthy farms over small farmers, which has the result of driving many small farmers out of business and into deeper poverty. The negative environmental impacts are characterized in terms of pollution. Rachel Carson (1962) established concerns over agricultural pollution in the public's consciousness when she exposed the unintended consequences of synthetic pesticides on nontarget species in *Silent Spring*. These fears are also real and well documented, as was seen in the last section.

There are ways to anticipate these fears and reduce ethical tensions in designing new GE crops. For instance, to the greatest extent possible, when designing GE crops, researchers should expand the goals beyond narrow productionism to include increasing social justice and reducing environmental impacts. Among other things, this will require reforming the ways agricultural biotechnology research is funded and incentivized. The key issues of an ethical and political debate over GE crops should be about how research is conducted, funded and incentivized to produce innovations that will contribute to social and environmental benefits, as well as production.

In an essay on environmental ethics and crop biotechnology, Paul B. Thompson (2003) offers two examples of research into GE crops that might reduce the ethical tensions over agricultural biotechnology. The first example involves farmers in the tropics who frequently save insect carcasses and pulverize them to spread on their crops. To some extent, the insect carcasses serve as an insecticide by spreading a virus that kills certain pests. The virus is naturally occurring and targets the insects causing problems for the farmers in the region. If scientists could identify the DNA sequence of the virus and insert it into crops traditionally used in this region, the new GE crops would mimic the crushed insects but be much more effective and require much less work. Furthermore, this proposed GE crop containing DNA from the virus would be a more sustainable and environmental benign alternative to chemical pesticides, which tropical farmers are rapidly adopting.

The second example is virus-resistant GE papaya. Papaya ringspot virus devastated the papaya industry in Hawaii in the 1990s. In response, growers widely adopted a virus-resistant GE papaya that has effectively addressed this problem. This technology does not necessarily favor large growers over small growers. Many poor people in developing countries grow papaya for household consumption; it is highly nutritious and plays a key role in a healthy diet. It costs no more to grow GE virus-resistant papaya than the natural variety. Further, the GE fruit does not require any changes to traditional growing methods. The argument here is that the virus-resistant GE papaya supports ecologically sound, small-scale farming practices. The purpose of briefly pointing to these examples, of which many more could be found if space allowed, is to reinforce the point that the polarized debate between techno-optimists and techno-pessimists is focused on the wrong issues. The ethical issues that we should be debating involve finding new strategies to guide research and funding mechanisms and public policies to support them.

Conclusion

In their recent book, *Big World, Small Planet*, Rockström and Klum observe that

> Agriculture today represents the single largest cause of biodiversity loss and greenhouse gas emissions (about 30 percent of global GHG emissions originate from agricultural production, roughly half from cultivation and the other half from deforestation). It is also the world's largest user of land (almost 40 percent of the world's terrestrial surface is under agriculture), and the largest user of freshwater (70 percent of the withdrawals of freshwater from rivers are used for irrigation). In addition, agriculture is the main source of nutrient overload from leakage of nitrogen and phosphorus into our water ways. (2016, pp. 167–168)

Once again, in order to correct these issues, agricultural practices will have to be transformed by finding a way to move beyond productionism and the technological fix strategy. Significantly, Rockström and Klum emphasize 'nature-based solutions' to create sustainable agricultural systems. For example, they discuss a poor region of Nigeria where farmers have combined nitrogen-fixing trees with crops in an agroforestry system to reclaim hundreds of thousands of hectares of degraded land and to increase yields. Societies must find ways to reward and incentivize these types of low-tech, high-intelligence approaches. However, Rockström and Klum's pragmatism philosophy of technology helps them avoid the ideological traps in the polarized GE debate. Along with nature-based approaches, there are potentially many important roles for agricultural biotechnology. Rockström and Klum note:

> Modern biotechnology can play a critical role in this regard. Combining genes into attractive combinations of healthy, resilient, sustainable, and productive food crops and species, without spreading to wild species or domestic equivalents, could be an essential part of the solution. Here we see the emergence of interesting potentials, moving away from the first generation of genetically modified organisms

(GMOs) associated with company dependence and uncertain side effects, to step-changes toward sustainable and productive foods, from crops to fish. (2016, p. 135)

Rockström and Klum's discussion demonstrates a willingness to bring seemingly opposing problem-solving approaches into dialogue. Their pragmatism illustrates the main goal of this chapter: to argue for technological pragmatism as a way forward in the polarized GE debate.

By framing the debate as pro-GE and anti-GE, techno-optimism and techno-pessimism are heightening tensions and missing the point. The important issues involve identifying limitations and defects in the productionist paradigm and technological fix strategy, then to take this knowledge and develop paradigm(s) and strategies capable of meeting the wicked problems of 21st-century agriculture. As a philosophy, technological pragmatism promotes dialogue among philosophers, scientists and policy makers to resolve the ethical tensions created by agricultural biotechnology. Paul B. Thompson's technological pragmatism sees the goal for agricultural ethics as an aid to helping 'build networks of communities that work to building ecologically sound and sustainable agriculture' (2003, p. 213). There are immense obstacles to realizing this goal. On one side, there is institutional inertia (educational, financial, cultural) built into research institutions that resists fundamental change. On the other side, there is inertia in environmental thought that has difficulty including agriculture in an ethical system at the scale needed to feed the over ten billion people who will live on the earth by 2100. However, the way to meet these challenges is through pragmatic engagement. Ideological purity only leads to polarization and heightens ethical tensions.

References

Borlaug, N.E. (2000) Ending world hunger: The promise of biotechnology and the threat of antiscience zealotry. *Plant Physiology* 124, 487–490. doi:10.1104/pp.124.2.487

Carson, R. (1962) *Silent Spring*. Houghton Mifflin, New York.

Cochrane, W.W. (1958) *Farm Prices: Myth and Reality*. University of Minnesota Press, St Paul, Minnesota.

Crabill, A., Sen, M. and Hochschild, J. (2012) Technology optimism or pessimism: How trust in science shapes policy attitudes toward genomic science. *Issues in Technology Innovation* 21. Available at: https://www.brookings.edu/research/technology-optimism-or-pessimism-how-trust-in-science-shapes-policy-attitudes-toward-genomic-science/ (accessed 24 April 2017).

Evenson, R.E. and Gollen, D. (2003) Assessing the impact of the green revolution, 1960 to 2000. *Science* 300, 758–762. doi:10.1126/science.1078710

FAO (2002) Unlocking the water potential of agriculture. Available at: http://www.fao.org/docrep/006/y4525e/y4525e04.htm (accessed 5 January 2018).

FAO (2004) *The State of Food and Agriculture 2003–2004: Agricultural Biotechnology: Meeting the Needs of the Poor?* Available at: http://www.fao.org/docrep/006/Y5160E/Y5160E00.htm (accessed 10 January 2018).

Foreman, D. (1991) *Confessions of an Eco-Warrior*. Harmony Books, New York.

Greenpeace (2013) Golden Rice. Available at: http://www.greenpeace.org/international/en/campaigns/agriculture/problem/Greenpeace-and-Golden-Rice/ (accessed 11 November 2017).

Kuhn, T. (1996) *The Structure of Scientific Revolutions*, 3rd edn. University of Chicago Press, Chicago, Illinois.

Lee, K. (1999) *The Natural and the Artefactual: The Implications of Deep Science and Deep Technology for Environmental Philosophy*. Lexington Books, Lanham, Maryland.

Levins, R.A. and Cochrane, W.W. (1996) The treadmill revisited. *Land Economics* 72, 550–553. doi:10.2307/3146915

Marx, L. (1983) Are science and society going in the same direction? *Science, Technology and Human Values* 8, 6–9. doi:10.1177/016224398300800402

Organic Consumers Association (2004) An open letter to Mr. Jacques Diouf, Director General of FAO, The FAO declares war on farmers not on hunger. Available at: https://www.organicconsumers.org/old_articles/ge/critic061604.php (accessed 7 January 2018).

Paarlberg, R. (2005) From the green revolution to the gene revolution. *Environment: Science and Policy for Sustainable Development* 47, 38–40. doi:10.3200/ENVT.47.1.38-41

Pence, G. (2002) *Designer Foods, Mutant Harvest or Breadbasket to the World?* Rowman and Littlefield, Lanham, Maryland.

Preston, C. (2008) Synthetic biology: Drawing a line in Darwin's sand. *Environmental Values* 17, 23–39. Available at: http://www.jstor.org/stable/30302622 (accessed 5 July 2018).

Rittel, H.W.J. and Webber, M.W. (1973) Dilemmas in a general theory of planning. *Policy Sciences* 4, 155–169. doi:10.1007/BF01405730

Rockström, J. and Klum, M. (2016) *Big World, Small Planet: Abundance Within Planetary Boundaries*. Yale University Press, New Haven, Connecticut.

Rosset, P. (2002) Taking seriously the claim that genetic engineering could end world hunger: A critical analysis. In: Baileym, B. and Lapppé, M. (eds) *Engineering the Farm, Ethical and Social Aspects of Agricultural Biotechnology*. Island Press, Washington, DC, pp. 81–93.

Sachs, J. (2005) *The End of Poverty, Economic Possibilities for Our Times*. Penguin Books, London.

Scott, D. (2011) The technological fix criticisms and the agricultural biotechnology debate. *Journal of Agricultural and Environmental Ethics* 24, 207–226. doi:10.1007/s10806-010-9253-7

Shepard, P. (1998) *Coming Home to the Pleistocene*. Island Press/Shearwater Books, Washington, DC.

Tenner, E. (1997) *Why Things Bite Back: The Revenge of Unintended Consequences*. Vintage Books, New York.

Thompson, P.B. (1995) *Spirit of the Soil: Agriculture and Environmental Ethics*. Routledge, New York.

Thompson, P.B. (2003) The environmental ethics case for crop biotechnology, putting science back into practice. In: Light, A. and DeShalit, A. (eds) *Moral and Practical Reasoning in Environmental Theory and Practice. MIT Press*, Cambridge, Massachusetts, pp. 187–218.

Thompson, P.B. (2009) Can agricultural biotechnology help the poor? The answer is yes, but with qualifications. *Science Progress*, 8 June. Available at: http://scienceprogress.org/2009/06/ag-biotech-thompson/ (accessed 25 May 2016).

Thompson, P.B. (2014) The GMO quandary and what it means for social philosophy. *Social Philosophy Today* 30, 7–27. doi:10.5840/socphiltoday201461210

Thompson, R.P. (2011) *Agro-Technology: A Philosophical Introduction*. Cambridge University Press, Cambridge, UK.

Tilman, D. (1998) The greening of the green revolution. *Nature* 396, 211–212. doi:10.1038/24254

Tilman, D., Cassman, K.G., Matson, P.A., Naylor, R. and Polasky, S. (2002) Agricultural sustainability and intensive production practices. *Nature* 418, 671–677. doi:10.1038/nature01014

Trewavas, A. (2001) The population/biodiversity paradox, agricultural efficiency to save wilderness. *Plant Physiology* 125, 174–179. doi:10.1104/pp.125.1.174

Trewavas, A. (2002) GM food is the best option we have. In: Pence, G. (ed.) *The Ethics of Food: A Reader for the Twenty-first Century*. Rowman & Littlefield, Lanham, Maryland, pp. 148–155.

Weinberg, A.M. (1969) *Reflections on Big Science*. MIT Press, Cambridge, Massachusetts.

Weinberg, A.M. (1972) Social institutions and nuclear energy. *Science* 177, 27–34. doi:10.1126/science.177.4043.27

Weinberg, A.M. (1994) *The First Nuclear Era: The Life and Times of a Technological Fixer*. American Institute of Physics, Woodbury, New York.

Weinberg, A.M. (2001) Can technology replace social engineering? In: Light, A., Katz, E. and Thompson, W.B. (eds) *Controlling Technology: Contemporary Issues*, 2nd edn. Prometheus Books, Amherst, New York, pp. 109–118.

White, L.T. (1967) The historical roots of our ecologic crisis. *Science* 155(3767), 1203–1207. Available at: http://www.jstor.org/stable/1720120 (accessed 5 July 2018).

Winner, L. (2004) Technology as forms of life. In: Kaplan, D.M. (ed.) *Readings in the Philosophy of Technology*. Rowman & Littlefield, Lanham, Maryland, pp. 103–113.

9 Absolute Hogwash: Assemblage and the New Breed of Animal Biotechnology

Katie M. MacDonald*

Department of Sociology and Anthropology, University of Guelph, Guelph, Ontario, Canada

Introduction

Intensive animal agriculture production sites present a number of logistical challenges. One of the most pressing issues surrounding these sites is the sheer amount of feces and liquid waste produced by large herds. Swine production in particular has been at the receiving end of much criticism surrounding the social and environmental risk of concentrated hog waste. Aside from being extremely odorous, hog excrement from commercial operations contains excess phosphorus. While phosphorus is needed to promote normal growth and function in hogs, a majority of the phosphorus available in commercial corn and soy-based feed contains a form of phosphorus called phytate-p. Hogs lack the enzyme phytase needed to digest phytate-p. As a result, phosphorus begins to build up in the digestive system, which is subsequently excreted.

In the late 1990s, researchers at the University of Guelph aimed to produce a bioengineered hog with a reduced output of phosphorus in its feces called the Enviropig (Golovan *et al.*, 2001; Forsberg *et al.*, 2013). This new breed of biotechnology emerged as a purported solution to increasing public unrest about elevated phosphorus levels and occurrence of eutrophication in rivers, lakes and water tables near industrial hog barns. Eutrophication causes dissolved oxygen

levels to deplete in aquatic bodies and can cause mass mortality, or dead zones, of plant and animal life (Correll, 1998). The Enviropig was developed by integrating a transgene comprised of a strain of *E.coli* phytase from a mouse. The addition of this transgene allowed the bioengineered hogs to secrete phytase in their saliva, allowing improved digestibility of phytate-p from commercial feed blends. As a result, there was a reduction in the concentration of phosphorus in feces (Golovan *et al.*, 2001; Forsberg *et al.*, 2013). However, the Enviropig was starkly different from other attempts at bioengineering as it was developed with the intention of being supplanted into the food supply as an environmentally conscious meat source. While the benefits of Enviropig addressed production issues, the hog from a consumption perspective was developed to be something that would line our grocery stores' deli aisles and thus become much more intimate to the average consumer. It was expected that consumers of meat concerned about the environmental impact of large-scale animal agriculture would be particularly drawn to this new form of genetically modified hog.

Drawing on Sanderson's (2015) notion of industrial hog production as assemblage, this chapter will profile the current structure of modern pork production to make a case for why ethical tensions in the food supply must be considered, but also why

* E-mail: kmacdo08@uoguelph.ca

their form is ever-evolving in light of new technologies. In the case of industrial hogs, the advent of new technology does not mean the old technology is obsolete. Rather, it creates a potentially contentious space for both early adopters and those refusing to adopt the advancements because it forces us to think about the replaceability of components of larger systems and how such components relate to others within the system. The Enviropig is unique as it also profiles the appropriateness of timing in technological adoption. The endeavor to engineer an animal with the genes of a mouse has frequently been purported to be 'before its time'. The role of transgenic, bioengineered and genetically modified organisms within agricultural systems destined for human consumption will undoubtedly increase as rising global temperatures, volatile oil prices, and severe and erratic weather threaten to dismantle and reshape our current productivist global food system. There are a number of ethical tensions in play here that this chapter will explore. These include whether undesirable aspects of animal agriculture can be technologically altered independent of their interactions with other aspects of the system and how the animal itself has become the new technology, particularly within hog production as animals are being developed to meet production needs. Because an ethical tension develops from contention or conflict, animal-based food systems are a particularly contentious issue in and of themselves, even devoid of biotechnology. Biotechnology, while not new, seems particularly foreign in relation to food animals.

Licensing Enviropig

The Ontario Pork Producers' Marketing Board, or Ontario Pork, was the patent licensee for Enviropig (Sanderson, 2015). Ontario Pork is the provincial marketing board for all hog producers in the province. The marketing board was formed in 1946 following a producer-led push to better organize production practices within the Ontario hog sector. Until 2010, Ontario Pork served as a single-desk seller for hog producers in the province which included negotiating hog carcass pricing contracts with processing plants on producers' behalf. Since the dissolution of single-desk selling, producers must directly negotiate their own contracts with plants, and Ontario Pork has shifted its focus toward consumer outreach, offering educational recipe cards, a blog, and increased governmental and retail representation.

Unable to find a major investor to continue funding Enviropig, the project was dismantled. All engineered hogs were terminated in May 2012 amid sharp criticisms from animal rights groups regarding the callousness of euthanizing ten otherwise healthy hogs. While corporate agriculture and the consumer base they deeply rely on may not have been ready for the bioengineered hog, biotechnology does and will continue to play a tremendously influential role in animal-based agricultural systems.

Assemblage: Tension of Taxonomy Made Unpredictable

Assemblage theory, first partially constructed by postmodernists Deleuze and Guattari in their 1987 book, *A Thousand Plateaus*, was later refined by the social philosopher De Landa (2006) to provide a framework for how humans ontologically affirm what things we perceive to legitimately exist and how components within a system change or evolve and relate to other components or aspects of systems. De Landa (2006) sought to operationalize elements of assemblage theory that were largely splintered throughout *A Thousand Plateaus*. To avoid a reductionist argument between abstractions of agency, individualism, and the construction and role of social structures, De Landa aimed to operationalize a more pragmatic theory of assemblage. De Landa notes that assemblages are 'wholes whose properties emerge from the interactions between parts' (2006, p. 4). While this is indeed a simplification of Deleuze and Guattari and later De Landa's work on assemblage, it allows us to consider the ethical impacts of bioengineering as series of interconnected or assembled pieces.

True to its postmodernist perspective, assemblage theory in biological and taxonomic consideration can serve to rupture and blur the properties and parts that we currently acknowledge to exist within animals as a whole. Under an assemblage framework, we are forced to think more systematically by considering how components interact, how replaceable components are, and whether and how replaceability affects other components within the system. Within this framework, we are able to better understand that borders can be cracked and weakened, even within the biological via bioengineering. Thus, in the case of an animal's biological function, each of the parts that make it a whole can be manipulated, extracted and repurposed to ultimately impact and emerge as a new whole. Is this appropriate and without consequence? Assemblage theory forces us to consider this issue.

Sanderson (2015) builds on De Landa's work by framing his piece as Enviropig-as-assemblage. Sanderson notes that using the Enviropig-as-assemblage framework allows for consideration of each of the actants or pieces that have amalgamated to create the Enviropig. These actants include the genetic sequencing used to supplant *E.coli* phytase from a mouse, the phrasing in legal documents used in the patenting of Enviropig and the controversial dialogue surrounding its use in the food supply (Sanderson, 2015). The assertion of Enviropig-as-assemblage serves as a conceptual framework to approach ethical rigidities within organisms, our food supply and technology, and forces us to think about the ethical impacts of altering a component that is part of a greater system.

More suited to this chapter, the notion of an assemblage – that the whole is comprised of the interactions between components – allows us to consider ethical tensions more clearly. Rather than reducing technological and biological changes as linear or evolutionary, we are forced to acknowledge our anxiety regarding the conditions for a bifurcated production system. One small change in the genome produces a huge, resounding impact that reverberates throughout the greater production system. Such a series of impacts will also surface further down the commodity chain as consumers will be forced to decide whether they prefer to consume conventional but environmentally damaging pork, or the bioengineered variety that is more environmentally friendly but has the genetic material of mouse woven in. However, while bioengineering creates ethical tensions that are best considered in assemblage, new technological advancements in hog production do not directly mean the old technology is obsolete because conventionally produced hogs will continue to be bred, grown and slaughtered along with the genetically modified variety. But now we must re-evaluate our perceptions of what is technology. Because the animal itself *is* the technology, additional discomfort and tension is experienced as our juxtaposition of *the biological* and *the technological* too is blurred. Decades of non-bioengineered genetic improvement, advancements in feed and veterinary care, housing and transportation equipment have resulted in the modern meat hog; the technology is the living organism. Using assemblage theory, each component, whether within the animal or those factors external to it, must be considered, which assemblage framework permits.

Much Too Much, Much Too Early?

As noted, one form of ethical tension develops from a discrepancy between old and new technologies. New technologies often replace older, existing technology creating an imbalance or rift between early adopters and those resistant to replacing existing technology. In the case of the Enviropig, not only does the animal itself become the technology, but the new technology does not necessarily mean that the existing technology is obsolete. Hog production in most of North America and increasingly abroad is done completely indoors; the barns, farm equipment and labor used toward growing a GM or conventional hog would be largely identical. But because the animal itself is the technology, rapid adoption presents a potentially contentious space for adopters/non-adopters due to anxiety arising from a perceived plurality of outcomes, since two similar yet different agricultural industries will operate simultaneously – one conventional and the other around the animal as technology. In fact, we are already seeing such splits occur in animal-based agriculture. In hogs, beef, egg and poultry production, we see two divergent industries operating simultaneously – one conventional and the other for animals that are free-range, cage-free, pasture-raised or housed on loose bedding. Biotechnology creates an additional cleavage within the system.

While dated, an article on technological adoption originally published in 1957 by the Cooperative Extension Service at Iowa State University, attempts to conceptualize how complex adopting new technology is in relation to farm practices and the type and quality of information sought/received about that new practice (Beal *et al.*, 1957, 1981). Beal, Rogers and Bohlen developed a theoretical framework of two overlapping scales to conceptualize how new agricultural technology is conceived of and how an iterative process of knowledge acquisition and reflection can shape one's readiness of adoption. The first scale considers five stages toward the diffusion of technology, which are: Awareness, Information, Application, Trial, Adoption. During Awareness, the producer is merely exposed to the idea but not sufficiently motivated to seek further information. The next stage of Information sees increased motivation to understand the new technology or practice and the individual uses information to compare and juxtapose this information to lived experiences. From here, Application can occur. During this stage, how this new technology or practice could be applied to one's existing situation is explored. Further reflection is required to decide about whether or not to try the new practice. Fourth, if the decision has been made to try the new practice, Trial is the stage where a producer attempts to apply the specific parameters of the technology to his/her situation. Lastly, Adoption occurs at the last stage of the thought process, resulting in a satisfied assessment and reflection that results in continued use of the technology. Acknowledging that these stages serve more as a conceptual scaffold than in absolute succession, the authors note that they

are useful for considering the complexity of technological adoption, particularly for agricultural producers.

In addition to the five-stage process above, Beal *et al.* aim to make further generalizations about how information, technology and practices are shaped and *timed*. Looking at hybrid seed corn, the authors provide a *diffusion curve* to understand the adoption rate for a new agricultural technology. The authors note that if data were illustrated as a simple distribution curve, there would be a normal, bell-shaped distribution across the stages of the diffusion curve. In order, the stages of the diffusion curve are: Innovators, Early Adopters, Early Majority, Majority, Non-Adopters, I will briefly outline each of these stages before linking to the argument that the rapid adoption of technology presents a place where both winners and losers are created, particularly for the Enviropig.

Innovators are those who first adopt a new idea or technology. Because of the increased risk of adopting new methods, innovators are often in a position of power within the industry or community and tend to have higher levels of capital and thus also larger production sites or farms. Unlike innovators, early adopters tend to have a slower rate of adoption to the new technology. These people tend to be well versed on industry happenings and are also well connected to industry (Beal *et al.*, 1957, 1981).

In contrast to the first two stages, once the early majority group adopts a change, widespread adoption increases rapidly. This group tends to be less active in industry or farm associations, but highly active in their community. This groups aims to be certain that the new technology will work before choosing to adopt it. The majority group, as the name implies, comprises the largest number of members. This group has not sought to educate themselves on the changes within the industry or engage in industry or community organizations, unlike previous groups. At this stage, the technological change or practice has largely become the new dominant method or approach. Finally, the non-adopters group consists of those who tend to be least educated on transitions within their industry, most conservative in their approach and therefore the most apprehensive about change or the uptake of new technology (Beal *et al.*, 1957, 1981). Having provided an overview of how diffusion of agricultural technology has been conceptualized historically, because the Enviropig has created conditions for the continued creation of winners and losers, there are two main paths that independent producers could have followed: those innovators who failed or those who would have been categorized as non-adopters and still largely failed.

First, innovators working on the Enviropig saw a potential to address production issues by developing what was claimed as a more environmentally friendly strain of pork. Because of increased public dismay of this bioengineered food animal and the inability to find a corporate investor to continue research, development and marketing of Enviropig, the project was dissolved. This new technology was slated to become the benchmark of how industrial animal agriculture could be engaged in a more sustainable manner, at least for water supplies. Because this innovative approach sought to develop a more water-friendly way of producing pork, Enviropig very well could have emerged as a monumental game changer, one that marked a huge pivot point in the hog industry. Early adopters could have seen increased profits as more producers opted to stock their barns with this improved technology. Processors and retailers could have seen a boom in profits as well as crafted advertisement campaigns and new value-added items could purport their wares as sustainable, possibly re-attracting those choosing not to consume pork out of environmental concern to once again eat meat. However, as we have seen and as will continue to be explored, the new technology failed.

Second, since this new technology did not directly compete with the old technology that is currently in use in industrial hog barns, those who could be categorized as non-adopters in line with Beal, Rogers and Bohlen's (1957, 1981) theoretical framework, even if Enviropig were to have progressed into mass production, would have still largely failed. More simply stated, those

independent producers who are currently reliant on old technology in the style of conventional production are at the whims of a commodity chain that is highly volatile, except for those positioned at the very top, such as processors and retailers. Processors and retailers hold sufficient influence and economic information pertinent to the broader pork industry, and are able to act accordingly and exploit their positions in the hog chain. In Canada, while domestic pork consumption is declining, nearly 70% of all processed meats in Canada are made with pork (Agriculture and Agri-Food Canada, 2015). The pork industry under the existent technology has indeed flourished, though independent producers are not the ones reaping the economic benefits. Because the Enviropig has created conditions for the continued creation of winners and losers, non-adopters would still have largely failed under the current technology.

Enviropig as Assemblage

Sanderson (2015) refers to the trajectory of the Enviropig as an *assemblage* because of its many interwoven conflicts and eventual failures. Sanderson further adds how the processes of patenting Enviropig added increased legitimacy and leveraging ability to the project by providing legal licensing and protection to this new type of invention. This assemblage is composed of a number of human actors and non-human actants, which collectively serve to make discourse heterogeneous or homogenous. If we consider the manure, we must also consider the pig which it came from, the feed formulated for its growth as well as the funding and labor necessary for each stage's conceptualization and operation. While the process of patent acquisition as one component of this assemblage does not literally give life to the animal, Sanderson states that this process provides the stability and instability needed to progress the life cycle of a bioengineered animal.

A bioengineered animal is quite different from a bioengineered crop and there are a number of documents which outline their management.

In 2003, the Cartagena Protocol on Biosafety to the Convention on Biological Diversity, an international agreement between 170 countries, was put in place to protect existing biological diversity from living modified organisms (LMOs) (Convention on Biological Diversity, 2017). The Cartagena Protocol aims to take a precautionary principle approach to LMOs in accordance with the principles established in the Rio Declaration on Environment and Development in an attempt to preserve and protect non-GM plants and animals, especially from any transboundary movement of LMOs and any adverse effects that ensue as a result.

Echoing the Cartagena Protocol's concern for GM animals' possible interference with existent animal stocks, the United States Food and Drug Administration (FDA) is presently soliciting commentary for their regulatory document concerning animals with 'intentionally altered genomic DNA' (FDA, 2017). What is interesting is that the FDA's definition is intended to be comprehensive in scope, in so far as to classify any alteration to animal DNA as a drug. The draft document states that, 'Unless otherwise excluded … the altered genomic DNA in an animal is a drug within the meaning of section 201(g) of the FD&C Act because such altered DNA is an article intended to affect the structure or function of the body of the animal, and, in some cases, intended for use in the diagnosis, cure, mitigation, treatment, or prevention of disease in the animal' (FDA, 2017, p. 7). The FDA's classification does not, however, extend to transgenic non-food animal species. These two documents illustrate two main concerns for biotechnology and its place within food animals. First, there is concern in relation to how a transgenic animal could impact the ecosystem if it is unable to be contained. Second, there is concern for what impact, if any, there is to human health when food animals are bioengineered and then consumed. It is both the notion of assemblage and the ability of Enviropig to somehow be supplanted outside of the isolated food system of which they are a part that causes increased tension towards the adoption of engineered food animals. While nation states

can place regulatory control on transgenic animals, as Sanderson notes (2015), patenting animals allows for increased flexibility and legitimacy of the product. Sanderson states that the creators of Enviropig sought simultaneous patent protection in a range of countries, before eventually receiving patents in China and the USA.

In the case of industrial hog production, high phosphorus levels were conceptualized as the primary problem. And while this is true, we must necessarily consider that the feed fed to hogs is largely indigestible and the massive scale at which these animals are being turned out for the industry's production needs compounds the environmental impact of phosphorus-laced feces. That is, high phosphorus levels in pig manure are a symptom of a larger, more problematic and complex food system. To draw on Sanderson (2015), by focusing on the GM as the problem we run the risk of isolating or silo-ing components within agriculture, all of which are a part of a much larger assemblage. Stated simply, the agricultural system is the problem, not the technology engineered to solve one aspect of it. Thus, if there is an advantage to the Enviropig saga, it is that we are forced to confront that reality of our modern food system as an *assemblage* as well.

Ethical Tensions of Biotechnology

Biotechnology in food animals is bioengineering made manifest. A bioengineered animal that can move under its own accord seems something eerily different from plant matter that is embedded in the earth or that is subjected to the whims of nature (e.g. when GM pollen is blown in the wind). It is *made*; literally designed out of flesh and blood. This is something much more distinctly wild and less controllable. The once whole assemblage of a hog was at one point indivisible. The whole could not function without the interplay of its components. Biotechnology has permitted the rupturing of the very order and role of biology. Hierarchical predictability and reliance on the linear is no longer the case. Perhaps we approach the assemblage of Enviropig with such ethical conundrum as we are forced to question what else in our lives has been fabricated to suit not just our wants, but what we are *told* we want. As advertisers have long known, connecting with a prospective customer on a personal level allows for a deeper connection and eventual implied trust in a company's brand. Once brand loyalty is established, it is easier to further sell new and innovative items. With Enviropig, perhaps consumer discomfort stems from the physical manipulation of flesh and blood, causing a crisis of intimacy that the altered flesh, once consumed, may continue to alter the flesh of the eater. One of the strongest threads of contention in the anti-GMO movement is that while selective breeding in crops and animal husbandry has been taking place for thousands of years since the onset of agriculture and the domestication of food animals, GMOs and any bodily reaction, adverse or otherwise, to their consumption, is not known well enough, particularly because of a lack of longitudinal data. While proponents of GMOs are quick to point to trials indicating no adverse reaction, avoidance of GMO products or inputs is met with a strong ethical conviction, despite current literature on their safety.

However, in the case of food animals, especially those for mass consumption, this is where we have misled ourselves. We have assumed that once an animal is bioengineered it becomes something bizarre and other. This new animal is still once a part of what it used to be, but still different enough to be seen as something abhorrent and unnatural. Bioengineering, by definition, allows for the optimal selection of favorable traits directed at a particular outcome; it is like we have pressed fast forward on evolution or rather, like we have by-passed evolutionary selection all together in favor of a being made to our own liking, needs and structure. We further attribute our unease about this new thing, now seemingly lacking naturalness, back to the idea that the animal has been developed for a specific use. In the case of Enviropig, the use was for production needs to address public outcry regarding one component of the hog assemblage: concentrated phosphorus in feces. It can be seen that in

such a scenario, the animal itself becomes commoditized. While it can certainly be argued that our industrial food system is based on the commodification of food, bioengineering and the ability to patent life literally results in an *animal as crafted commodity*. Here, the animal has been shaped, manipulated and formed into a specimen that unequivocally suits our needs.

However, these developments are already happening. Our fear of the lab-developed animal, the animal that is perceived as so foreign and other, already exists. For hogs, the breeds being utilized for commodity/commercial/industrial production have resulted from millions of dollars in genetic research. Such genetic advancements are not *bioengineered* as in the case of the Enviropig hosting the genetic material of a mouse resulting in a truly transgenic animal, but are engineered nonetheless. The substantial amount of capital infused into genetic research and development does mark an accelerated push with regards to animal biology. These genetic advancements which seek improved foot hardiness for growth over concrete floors, large litter sizes and carcass leanness are arguably still pushing the development of an animal that is no longer 'natural'. It seems that the contention of the Enviropig lies with the notion that the animal itself has become the technology. The ethical dilemma of how we substantiate the point of *too much technology* is blurred. In order to more accurately frame this argument, a brief overview of the breeding process of hogs must be provided. Similar scenarios exist for most food animals as dairy cows and poultry are also artificially inseminated, rely heavily on improved genetic strains and will too face criticism regarding the limits of biological function.

Current State of Hog Production

In Canada, as in the USA, much of industrial hog production is done indoors. Drabenstott (1998) notes that the US hog industry underwent a drastic shift in the 1980s, claiming that 'the new pork industry' was defined by three main characteristics that included a jump in the percentage of growth under contracts, a more concentrated industry overall and a dramatic change in the geography of production. Drabenstott (1998, p. 80) further notes that a cornerstone of this 'new pork industry' was further driven by ongoing improvements in both genetics and production techniques, and confined housing advancements allowing for increased production through economies of scale. Stull and Broadway (2004, p. 12) reiterate how 'agricultural industrialization' has also impacted and transformed meat production. Industrial meat production, including hogs, has been reshaped into an industry reliant on intensification, concentration and specialization. Due to the boom in growth of scale, these production sites have been aptly termed intensive feeding operations, confined animal feeding operations or intensive livestock operations.

As sites host larger and larger populations of hogs, there is also a trend towards multi-site production loops. Within multi-site production loops, the lifecycle of the hog is split across different sites. While some producers engage in fully integrated farrow-to-finish production, where producers oversee impregnation, farrowing and the growth of weaned piglets to market weight, multi-site production loops allow for increased specialization. Production sites can also include farrow-to-wean, nursery, wean-finish and feeder/finishing operations. What is most important to consider here is that regardless of the type of production site, industrial hogs are predominantly bred using artificial insemination. For example, in Ontario there are a number of companies (Fast-Genetics, EastGen, Genesus) that specialize in hog genetics, specifically providing producers with top-quality frozen boar semen for use on sows. These genetics companies also offer quality sow lines rather than have producers pull next generation breeders from existing stock. Industrial hog farms have become so entrenched in the margins of production that when sows are spent and sent off to slaughter, producers will often source a gilt (or female hog that has not yet given birth to a litter of piglets) from a genetics company to ensure she is equipped with the latest iteration of optimal genes,

rather than source potential sow herd from existing gilts born on-farm.

Once the gilt/sow has been artificially impregnated from boar semen that is also increasingly sourced frozen from genetics companies, she is housed within a gestation crate. There is an industry standard for gilts/sows to be induced to farrow following a 114-day gestation period (Ménard, 2015). At this stage, she is moved into a farrowing dry stall. These dry stalls, much like the gestation crates, limit the movement of the sow to avoid piglet crushing post-farrow, but are equipped with extra space for piglets, called a creep area. On commercial operations, it is typical for gilt/sow farrowing to be precisely scheduled. Due to the intensive labor requirements needed on-farm during and after farrowing, commercial production sites have trended towards batch-farrowing. Here, all gestating gilts/sows are induced and scheduled to farrow over the course of a chosen week rather than having gilts/sows sporadically farrowing (Ménard, 2015). Farrowing is a particularly labor-intensive process within hog production for three reasons.

First, typically within seven days of birth, piglets are given iron shots, their tails are docked to deter tail-biting in crowded pens, their needle/wolf teeth are clipped to avoid injury to the sow's teats and the piglet's littermates and male piglets are castrated to deter the development of boar taint, which develops from hormones released from the reproductive organs, resulting in a foul-smelling and tasting cut of pork post-slaughter (National Farm Animal Care Council, 2014).

Second, because of their restricted movement throughout the gestation period, gilts/sows fail to properly develop and strengthen the muscles that are needed to give birth. The result is that when a gilt/sow begins to farrow, many commercial sows' pelvic muscles are too weak to push the piglets through the birth canal. Manual farm labor needs to be present at the time of farrowing to essentially pull piglets out of the sow's body to avoid both the piglet and sow from perishing.

Third, due to genetic advancement, there is a tremendous industry push toward increased

sow productivity, which includes greater litter sizes (Brisson, 2014). Sows typically have
between 14 and 16 teats, yet genetics companies aim to produce breeding stock that
consistently births litters of 14–24 live piglets. Litter variation often occurs between
piglets and size difference can result in the
development of a teat order (National Farm
Animal Care Council, 2014; Ménard, 2015).
One's position in the teat order is imperative
when there are more mouths than teats.
Larger, stronger piglets often feed better
than smaller piglets, and as a result, piglets
are often reorganized into like-sized groups,
a practice called split-suckling (Ménard,
2015). When batch-farrowed, all sows will
be dropping milk at about the same time,
but may not be nursing the piglet that she
gave birth to.

By knowing exactly when on-farm labor
is needed, producers are also able to save on
labor costs and streamline operations, allowing for great control over the biological
phases of the hog lifecycle. But regardless of
the reader's stance or concern for animal
welfare, the point here is that the modern
food system, particularly those that are
animal-based, thrives on the exploitation of
biological limits of the physical animal for
corporate gain. Thus, the Enviropig serves as
a marvelous example of the exploration for
ethical tensions in food production animal, as
it can be seen as unnatural, problematic to
biodiversity and an unethical experiment in
genetic wizardry. However, the most apparent ethical tension stems not from the addition or alteration of genetic material, but
from the practice and normalization of developing, altering and maximizing the body of
the animal to meet seemingly insatiable consumer and production needs. Further, when
considered as assemblage, tensions and anxieties reaching beyond the barn force us to
consider our relationship to technology,
consumerism and the very way we frame
how we look at, interact with and understand our lived realities.

More specifically, modern industrial hog
production presents two major problems.
First, we have reduced issues to a singular
cause: If toxic phosphorus levels, in the case
of the Enviropig, were identified as the major

environmental condition caused by industrial hog farms, and if such high phosphorus
levels are found in the feces of commercial
hogs, then phosphorus levels *alone* must be
addressed. As mentioned, scale of production is central here. Phosphorus levels, while
still high when excreted from the individual
hog, are amplified when a production site
hosts tens of thousands of animals. Second,
we have reduced the solution to a singular
impact. We have compartmentalized the
problem and solution in isolation of the
greater system it is a part of. We have not
considered the pig as a part of an assemblage
as Sanderson (2015) suggests. Researchers
identified the low bioavailability of phosphorus via digestion as a significant reason for
the concentrated phosphorus levels in hog
feces. Perhaps feed formulas need to be reconsidered and reformulated to make phosphorus
more bioavailable. Perhaps hogs' historically
omnivorous diets, combined with high digestibility of phosphorus from animal sources
(like bone and animal meals), play a considerable factor in the attempt to derive plant-
based phosphorus intake needs from a largely
indigestible source. Reductionist problem
solving serves to further cloud tangible,
more holistic solutions towards system-wide
change.

In-Barn: Where New and Old Technologies Amalgamate

The pork that lines our supermarkets has
been systematically cleansed of that which is
uncomfortable. Industrial barns and their
accompanying foul-smelling air have been
pushed out to the fringes of rural communities, often employing transient, desperate
populations (Constance and Bonnano, 1999;
Grey, 2000; Novek, 2003). But for some consumers they are not far enough away. In
Canada, one of the largest slaughterhouses
in Ontario, Quality Meat Packers, was met
with constant resistance from a growing
urban population that expressed discomfort
at the sights, smells and sounds of slaughter
from their high-rise condominiums. Pachirat's (2011) work on industrial beef slaughter reiterates the notion of discomfort with

being too close to the death that is a necessary part of meat production, what he terms a politics of sight. However, with bioengineered animals intended for the food supply, we are faced with reality that our drive to consume will necessitate an even greater shift in production – in this case, a massive biologically crafted shift. Ethical tensions emerge from our recognition of animals as new technology as they are further developed through advanced scientific and technological means to meet production and consumption needs. We might accept that animals are raised in confinement, that much of the perceived naturalness and idyllic conceptions of natural farming have been removed in modern industrial food animals, and that many food animals are just a whisper or shadow of the animals they were domesticated from. Yet, there is discomfort at the addition of microscopic genetic material in animals, resulting in our being forced to consume that which we know is foreign. Our unease with commercial animal systems has caused us to push these sites out of view literally and figuratively push them from nature through genetic modification. Our cognitive dissonance has allowed us to be accepting of large scale, concentrated animal production sites. Yet, we villainize and scoff at the addition of genetic material claiming to better enhance the current production system.

There is also discomfort in knowing that our insatiable patterns of consumption have been met with an animal better equipped to slow environmental degradation. However, what is troublesome is that the Enviropig makes manifest the ills of our modern food supply – a food supply built on the continued creation of winners and losers. Perhaps it is not the discomfort with biotechnology specifically, but rather with the concern that we are not really in control of what we consume, resulting in an increasingly fragile and broken food system. But the Enviropig forces us to face the realities of our productivist food system. It may be easier to place blame on new, emergent technology than to recognize the reality that issues of animal agriculture continue to be contentious spaces as the new technology does not directly mean the old technology is obsolete, but that both approaches are deeply

flawed. The ethical quandary with bioengineering *per se* is an interesting one as we are forced to acknowledge the system or assemblage that industrial meat exists within. Rather than dismiss the applicability of technology made manifest, we must work to understand the many inputs, hands and nodes through which industrial meat must pass, the corporate-industrial regime that dictates both what is produced and consumed, and the abhorrent welfare, social and environmental costs which are indeed absolute hogwash.

References

Agriculture and Agri-Food Canada (2015) Canada's red meat and livestock industry at a glance... Government of Canada. Available at: http://www.agr.gc.ca/eng/industry-markets-and-trade/statistics-and-market-information/by-product-sector/red-meat-and-livestock/red-meat-and-livestock-market-information/industry-profile/?id=1415860000002 (accessed 2 December 2015).

Beal, G.M., Rogers, E.M. and Bohlem, J.M. (1957) Validity and the concept of stages in the adoption process. *Rural Sociology* 22, 166–168.

Beal, G.M., Rogers, E.M. and Bohlem, J.M. (1981) *The Diffusion Process*. Special Report No. 18: Cooperative Extension Service. Iowa State University, Ames, Iowa.

Brisson, Y. (2014) The changing face of Canada's hog industry. Catalogue no. 96-325-X – No. 005. Statistics Canada, Ottawa, Canada.

Constance, D.H. and Bonanno, A. (1999) CAFO controversy in the Texas panhandle region: The environmental crisis of hog production. *Culture & Agriculture* 21, 14–26. doi:10.1525/cag.1999.21.1.14

Convention on Biological Diversity (2017) *The Cartagena Protocol on Biosafety to the Convention on Biological Diversity*. United Nations Environment. Available at: https://bch.cbd.int/protocol/ (accessed 2 June 2017).

Correll, D. (1998) The role of phosphorus in the eutrophication of receiving waters: A review. *Journal of Environmental Quality* 27, 262–266. doi:10.2134/jeq1998.00472425002700020004x

Deleuze, G. and Guattari, F. (1987) *A Thousand Plateaus: Capitalism and Schizophrenia*. Athlone Press, London.

De Landa, M. (2006) *A New Philosophy of Society: Assemblage Theory and Social Complexity*. Continuum International Publishing Group, New York.

Drabenstott, M. (1998) This little piggy went to market: Will the new pork industry call the heartland home? *Economic Review–Federal Reserve Bank of Kansas* 83, 79–97.

FDA (2017) Guidance for Industry (GFI) #187: Regulation of Intentionally Altered Genomic DNA in Animals. Available at: https://www.fda.gov/downloads/AnimalVeterinary/GuidanceComplianceEnforcement/GuidanceforIndustry/UCM113903.pdf (accessed 2 June 2017).

Forsberg, C.W. *et al.* (2013) Integration, stability and expression of the *E. coli* phytase transgene in the Cassie line of Yorkshire Enviropig™. *Transgenic Research* 22, 379–389. doi:10.1007/s11248-012-9646-7

Golovan, S.P. *et al.* (2001) Pigs expressing salivary phytase produce low-phosphorus manure. *Nature Biotechnology* 19, 741–745. doi:10.1038/90788

Grey, M.A. (2000) 'Those bastards can go to hell!' Small-farmer resistance to vertical integration and concentration in the pork industry. *Human Organization* 59, 169–176. doi:10.17730/humo.59.2.0652121054761873

Ménard, J. (2015) Practical labour tips to improve piglet survivability. *Proceedings of the 2015 London Swine Conference: Production Technologies to Meet Market Demands*. London Swine Conference, London, Ontario, Canada.

National Farm Animal Care Council (2014) *Code of Practice for the Care and Handing of Pigs*. Canada Pork Council and National Farm Animal Care Council, Ottawa, Ontario, Canada.

Novek, J. (2003) Intensive livestock operations, disembedding, and community polarization in Manitoba. *Society and Natural Resources* 16, 567–581. doi:10.1080/08941920309188

Pachirat, T. (2011) *Every Twelve Seconds: Industrialized Slaughter and the Politics of Sight*. Yale University Press, New Haven, Connecticut.

Sanderson, J. (2015) Who killed the EnviroPig? Assemblages, genetically engineered animals and patents. *Griffith Law Review* 24, 244–265. doi:10.1080/10383441.2015.1063570

Stull, D.D. and Broadway, M.J. (2004) *Slaughterhouse Blues: The Meat and Poultry Industry in North America*. Thomson Wadsworth, Toronto, Ontario, Canada.

10

Nature-identical Outcomes, Artificial Processes: Governance of CRISPR/Cas Genome Editing as an Ethical Challenge

Frauke Pirscher,[1] Bartosz Bartkowski,[2]* Insa Theesfeld[1] and Johannes Timaeus[3]

[1]*Institute of Agricultural and Nutritional Sciences, Martin Luther University Halle-Wittenberg, Halle, Germany;* [2]*Department of Economics, UFZ – Helmholtz Centre for Environmental Research, Leipzig, Germany;* [3]*Department of Ecological Plant Protection, University of Kassel, Kassel, Germany*

Introduction

CRISPR/Cas is a newly developed genome editing technique that is viewed as revolutionary to crop breeding.[1] It allows for modifications of genes by adding, cutting out or suppressing certain gene sequences of the DNA. Compared with former genetic modification (GM) techniques, this system is considered relatively easy to apply, more precise, quicker and much cheaper (Baker, 2014). For example, according to Ledford (2015), the cost difference between applying CRISPR/Cas vs zinc finger nucleases (ZFN), another common genome editing technique, is in the range of two orders of magnitude (US$30 vs US$5000). Therefore, CRISPR/Cas is expected to have great innovative potential in agriculture by speeding up breeding, increasing yields and allowing plant production under less favorable conditions (Baltes *et al.*, 2017). It has already been successfully applied in breeding of different agricultural plants, such as soybean (Jacobs *et al.*, 2015; Li *et al.*, 2015), maize (Svitashev *et al.*, 2015), tobacco, sorghum, rice (Woo *et al.*, 2015) and tomato (Brooks *et al.*, 2014; Ueta *et al.*, 2017).

Although CRISPR/Cas generally allows for trangenesis, that is, transferring DNA sequences across species, its current application in agricultural biotechnology mainly remains within species boundaries, including transfers of genes between varieties (cisgenesis *sensu stricto*), allele replacement, gene knockout and gene silencing. In the following we use 'cisgenesis' as a generic term encompassing all those non-transgenic variants for simplicity. In other words, cisgenesis allows for creating new products that could also be the outcome of natural evolution or conventional breeding. A unique characteristic of CRISPR/Cas-based cisgenesis is that it does not leave detectable traces of genetic engineering (especially foreign DNA snippets) in the resulting organism, thus its results effectively cannot be identified as genetically modified organisms (GMOs): they are so-called nature-identical genetically modified organisms (nGMOs).[2] Since CRISPR/Cas is extremely quick and, in contrast to earlier genome editing techniques, cheap, it can be seriously

---

* Corresponding author. E-mail: bartosz.bartkowski@ufz.de

considered as an alternative to conventional breeding for some applications.

The former GM debate was dominated by discussions on risk assessment and general objections against the transgression of species boundaries (e.g. Gaskell *et al.*, 2010; Kvakkestad and Vatn, 2011). It was embedded in a general critique of modern agriculture, especially the property rights to genetic resources (e.g. patents on genetically modified plants). In the face of the differences between former GM and CRISPR/Cas technology, the question arises how this has shaped the societal debate on genetic engineering of plants, and how it will shape it in the future. Because nGMOs can be molecularly identical to conventionally bred or wild plants, the question arises of whether they should be viewed as GMOs at all. The decision regarding their classification has consequences for the general need and possible ways of governing genetically engineered products. The classification of CRISPR/Cas modified plants as GMO is more than a scientific and legal issue: governance is key to balancing diverging interests and values within a society and thus avoiding a loss of public trust in certain technological developments. In a democratic society, legitimacy is at the center of governance (Vatn, 2016; Meinard, 2017); in normatively sensitive contexts, such as genetic engineering, it is particularly important that governance constitutes an attempt to strike a balance between the diverging interests and values.

Against this background we provide a brief introduction to the CRISPR/Cas genome editing technology and introduce the critical concept of nature-identical GMOs. We then outline the current societal debate on regulating cisgenic plants created by means of CRISPR/Cas genome editing, with a particular focus on the European Union (EU). We discuss the different value concepts and ethical tensions that lead to different perspectives on regulation and frame possible governance solutions that may help to balance those diverging values by highlighting governance-specific challenges. Finally, we offer conclusions about how CRISPR/Cas genome editing gives rise to a set of ethical tensions that go well beyond previous debates surrounding GM crops, particularly by opening up a debate about a product vs process perspective – in both ethical and governance discourses.

CRISPR/Cas Genome Editing

A range of different GM technologies are referred to as genome editing (Baltes *et al.*, 2017). However, the CRISPR/Cas technology is the prime example of this type of biotechnology. It is closest to fulfilling the promises of the editing metaphor: changing genetic code as easily as editing a digital text document.

While CRISPR/Cas is mostly associated with biotechnology, it is actually based on a defense mechanism that occurs naturally in bacterial cells: if a bacterium survives the first attack of a virus, the cell integrates parts of the viral DNA in its own genome in the CRISPR sequence. The next time the virus infects the cell, the viral DNA copy serves as a targeting mechanism. This DNA sequence is transcribed into an RNA molecule that binds to the Cas9 enzyme and recognizes the DNA of the virus by base-pairing, 'guiding' the enzyme, which then cuts the viral DNA at a precisely defined sequence and thus neutralizes the virus. This precise and highly adaptive bacterial immune system is the natural foundation of the CRISPR/Cas technology.

In their seminal paper, Jinek *et al.* (2012) presented a new genome editing tool based on the natural CRISPR/Cas system.[3] They found that two RNAs (crRNA and tracrRNA) are required to guide the Cas9 enzyme to the target sequence and simplified the natural mechanism by fusing them into a single guide RNA. Their study delivered a GM technology that combines the strength of enzymes as precise gene scissors and the strength of RNA as precise targeting mechanism for specific DNA sequences that can be easily synthesized in the lab to target particular genes.

CRISPR/Cas is not the only genome editing tool allowing for precise alterations of DNA. However, earlier genome editing technologies combine the gene scissor and the targeting function in just one protein molecule. Therefore, for every gene one wishes to edit, a new protein has to be engineered so as to adapt the targeting part of the protein to the gene sequence. Engineering proteins is time-consuming and costly due to their complex chemical structure. Therefore, a profitable application of these technologies has proven difficult. The CRISPR/Cas tool separates the scissor function (protein) and the targeting function (RNA). Only the RNA needs to be modified for each new application. Because RNAs are much easier to engineer in the laboratory, this results in a much more efficient process.

Another novel aspect of CRISPR/Cas is that it makes so-called multiplexing much easier than other GM techniques (Cong et al., 2013): multiple guide RNAs, targeting different DNA sequences, can be introduced into a cell together with the Cas9 protein at once. Since plant genomes are often polyploid, that is, carrying multiple gene copies, genetic engineering of a single trait (i.e. characteristic of the organism) requires editing multiple DNA sequences. A common example is hexaploid wheat carrying six copies of each gene. Therefore, multiplexing is particularly useful for plant breeding (Khlestkina and Shumny, 2016).

While CRISPR/Cas does nothing on the molecular level that could not be done before, its efficiency and simplicity have profound consequences for what can be done to breed new crop plants and livestock varieties.

Nature-identical CRISPR/Cas Products

As already briefly mentioned, a potentially revolutionary aspect of CRISPR/Cas genome editing is that it makes possible the creation of nature-identical GMOs, that is, organisms that could just as well be the result of crossbreeding or spontaneous mutation (see also Bartkowski et al., 2017). When applied for cisgenesis (Cardi, 2016) – the transfer of genes between sexually compatible organisms (those that can also be crossbred) – or gene knockout (where a DNA sequence is cut out but not replaced by a new sequence), CRISPR/Cas is, contrary to earlier GM techniques, not supposed to leave any traces of the modification. This is because CRISPR/Cas genome editing uses the natural in-cell repair mechanisms to insert the wished-for DNA sequences in the DNA of the target organisms at the location cut by the CRISPR/Cas complex; the Cas protein and guide RNA themselves are degraded by the cell.

The non-traceability assumption has been challenged (Kim and Kim, 2016). However, at least when using common tests, CRISPR/Cas-generated modifications cannot be identified as such (Waltz, 2016). Furthermore, the possibility of off-target modifications has been considered a challenge for genome editing (Schaefer et al., 2017), even though they can potentially be removed by means of backcrossing (Belhaj et al., 2015). Both issues will be discussed further in the following sections.

Given that CRISPR/Cas genome editing is highly precise and quick, allows for multiplexing and is significantly less costly than other GM techniques, it is a serious alternative to conventional breeding in the context of breeding within species boundaries (including gene silencing and gene knockout), although it still requires more resources in terms of equipment, know-how, and so forth. Moreover, because genome editing changes only very specific DNA sequences, while crossbreeding almost always results in multiple changes and thus in new varieties, CRISPR/Cas products may be more attractive especially in traditional markets such as grape vines, where varieties that are very popular and have a long tradition could be maintained but enhanced in terms of resistance against new pest species. In contrast, the application of CRISPR/Cas requires knowledge of gene sequences responsible for the targeted trait. Currently, it cannot thus replace conventional breeding completely. Nonetheless, in the middle to long term, nGMOs can be expected to become quite widespread; this development would have profound moral consequences.

Societal Debate on CRISPR/Cas

The former GM debate has been dominated by fundamental ethical objections against modifications that violated species barriers, which a majority of EU citizens wanted to be respected; GM products have been viewed as unnatural. Besides these principle ethical objections, the debate has focused on risks of possible off-target effects caused by deliberate release of products with 'unnatural' gene constructs. Against this background, proponents of the application of genome editing now take up this product-oriented perspective to argue in favor of CRISPR/Cas. In the case of a cisgenic plant, the gene of interest, together with its promoter, has been present in the species or in a sexually compatible relative for centuries. Therefore, cisgenesis does not alter the gene pool of the recipient species and does not provide 'additional' traits. Proponents of genome editing conclude that no changes in fitness occur that could not happen through either traditional breeding or natural gene flow and that cisgenesis thus carries no risks – such as effects on non-target organisms or soil ecosystems, toxicity or possible allergy risks emanating from GM food or feed – other than those that are also incurred by traditional breeding. From an outcome-oriented perspective, the major criticisms against GM do not hold true anymore. Therefore, it is the hope of many GM advocates that CRISPR/Cas will solve the 'GM problem' by leading to higher general acceptance. Empirical studies of consumers' preferences show that cisgenesis is in fact viewed much less critically than transgenesis (Delwaide *et al.*, 2015; Edenbrandt *et al.*, 2018).

From this perspective it is therefore only logical to treat CRISPR/Cas modified plants like traditionally bred varieties in EU regulation. However, current EU law defines GMOs as 'organisms in which the genetic material (DNA) has been altered in a way that does not occur naturally by mating or natural recombination' (EU Directive 2001/18). Thus, according to the usual interpretation, GMOs are defined by the process of creation. The same is true for all international law except in the USA and Canada (Schouten *et al.*, 2006). The regulation does not differentiate between transgenic and cisgenic organisms. Thus, currently both types of organisms fall under the EU regulation. New

crop plant varieties that have been developed on the basis of genetic engineering have to undergo a procedure of approval that is more demanding than the conventional approval for food products. In the EU, products that contain more than 0.9% of GMO have to be labeled as GM products. At the same time, EU law allows for exceptions from this procedure in cases of mutagenesis and cell fusion (EU Directive 2001/18, Annex 1B).[4]

The possibility of cisgenic modification by CRISPR/Cas has led to calls for a revision of the current EU regulation of GM. Leading breeding companies (KWS, 2015), as well as a group of scientists in multiple recent opinion papers (Araki and Ishii, 2015; Huang *et al.*, 2016), argue for product-based rather than process-based regulation. Some call for adding cisgenesis to the list of exceptions of techniques for the GM regulations (Schouten *et al.*, 2006).

The way of regulating these new crop plant varieties has enormous consequences for the development costs of these crops. The cost of approval may exceed the cost reduction achieved by the breeding technique. The amount of upfront costs will also influence the market structure of the seed market. The key argument made here by proponents of a relaxation of the regulatory burden has at its core the molecular identity between the natural and modified plant that leads them to assertions that the molecular identity guarantees for no additional risk caused by the breeding.

Consequently, from the perspective of the proponents of law revision, the deliberate release and market introduction of cisgenic plants can be considered as safe as the release and market introduction of traditionally bred plants. On the issue of safety, regulators could treat cisgenic plants the same as conventionally bred plants (Schouten *et al.*, 2006).

The ongoing debate on a revision of the current GM law regulation, however, has shown that the shift in focus towards cisgenesis has not muted ethical concerns; rather, it has shifted them into another direction. Now, critics argue that although the product might be nature-identical, the process is highly artificial. All GM techniques directly interfere at the level of the genome by inserting material that was produced outside the cells. This degree of intrusion into the cell is viewed as risk in itself because the consequences for the plant are insufficiently known. Critical biologists like Then (2016) and Steinbrecher (2015) argue that CRISPR/Cas is indeed more precise than former GM techniques but is still not free of errors. Unintended alterations of the genome can occur because of insufficient specificity of the nuclease, so that the DNA will be cut in several regions of the genome instead of the one intended (see Schaefer *et al.*, 2017). This might happen irrespectively of the number of DNA pieces that should be inserted or their length. Thus, the technique might have off-target effects (Steinbrecher, 2015). Therefore, critics call for a regulation of the process, so as to guarantee for a traceability of the genetic modification, and for consumers' right to know which food products are produced from genome-edited plants. From their perspective, long-term safety of the technique cannot be reliably guaranteed up to now. From a process perspective a risk still exists, irrespective of the molecular equivalence of nGMOs. But there are also principle concerns, which are not totally new, but now gain much more importance. It is the general critique at human intrusion into natural processes, often framed as 'Playing God'. This argument does not become muted when CRISPR/Cas applications respect species barriers. With growing precision in genetic manipulation, the distinction between 'living' vs 'non-living', 'nature' vs 'artificial' is further blurred and can no longer be taken for granted (Ried *et al.*, 2011). Therefore, opponents of genome editing ask whether there is a red line that should not be crossed, how it is defined, and who decides about the need for and legitimacy of research. Thus, congruency of scientific and societal interests cannot be achieved by simply shifting from conventional GM to CRISPR/Cas genome editing. As long as there is no clear answer to these open questions, critical voices will consider process-based regulation as urgently needed and in line with the precautionary principle.

Similar divergences become even more apparent in the discussion about the acceptance of CRISPR/Cas for organic breeding. Previous applications of GM techniques have been oriented at the needs and problems of conventional farming. In organic farming the breeding goals are different, mainly because of the general rejection of synthetic fertilizer and pesticides. But the introduction of genes from older varieties offers a possibility to cope with the specific requirements in organic farming (Ronald and Adamchak, 2008). The application of CRISPR/Cas technique in breeding could speed up the breeding successes tremendously. When Urs Niggli, head of the Swiss Research Institute of Organic Agriculture Fibl, proposed CRISPR/Cas as a possible option in organic breeding (taz, 2016), the majority of organic farmers' associations dissociated from his position (Bio Suisse, 2016; Bioland, 2016). For them, the CRISPR/Cas technique is a genetic modification and therefore not compatible with the principles of organic breeding and farming.

In scientific literature there are different positions regarding the compatibility of organic farming with new breeding techniques. Generally, organic farming interprets agriculture as a process that includes agroecological, socioeconomic and ethical principles. IFOAM, the International Federation of Organic Agricultural Movements, has defined four principles to reflect this holistic understanding of agriculture: health, ecology, fairness and care (IFOAM, 2014). It is argued that new breeding techniques like CRISPR/Cas are not compatible with these principles and the underlying values of organic farming. Conflicts mainly occur with the principle of health, understood as supporting the wholeness and integrity of living systems. From this biocentric perspective, any alteration of the DNA violates the integrity of the genome as part of a living entity (Lammerts van Bueren *et al.*, 2003; Nuijten *et al.*, 2016). However, with regard to rewilding (i.e. the backward breeding of traits from old varieties or wild relatives), some argue that new breeding techniques should be accepted in organic agriculture (Andersen *et al.*, 2015; Palmgren *et al.*, 2015), because they can reduce the need for chemicals and backward breeding, but also because reverse breeding is always a step back to nature and the rewilded organisms are more natural than the source plant and therefore compatible with the principles of organic farming (Andersen et al., 2015; Palmgren et al., 2015). Some of those feeling affiliated to organic farming regard it as not only compatible with the principle of health but a very efficient way to promote this principle.[5] Proponents of the application of CRISPR/Cas in organic farming view labeling of new breeding technique's products as GMOs as an impediment for the implementation of reverse breeding, while the majority of representatives of organic farming call for a consequent risk assessment and labeling under GMO regulation to guarantee a non-GM production and transparency, and thus the freedom of choice for farmers and consumers.

CRISPR/Cas-modified Plants: Conflicting Values and Ethical Tensions

In addition to definite economic interests that guide the debate, different opinions about the need and acceptability of human interference in nature can lead to opposing positions and ethical tensions regarding the treatment of CRISPR/Cas modified plants. One should keep in mind that breeding is by definition human interference into natural processes, as it is the ultimate aim of breeding to change plants according to human needs. Humans have done this since the Neolithic Revolution. Nevertheless, breeding and agricultural production are not completely artificial, but depend on natural systems. However, discovering nature up to the molecular level, decoding and modifying natural processes in more detail has been further blurring the boundaries between nature and technology.

In the current public debate, naturalness is not only used for describing different degrees of intrusion into plants, but also as moral category (van Haperen et al., 2012). However, the classification 'natural' or 'unnatural' is ambiguous. The above outlined positions on calling for or rejecting regulation of cisgenic crops show that proponents and

opponents use different understandings of naturalness. In her article 'Dimensions of Naturalness', Siipi (2008) classifies different interpretations of naturalness or unnaturalness generally occurring in the public and scientific debate in bioethics. These categories are helpful to understand why it is now possible to argue both for and against CRISPR/Cas by referring to naturalness. The two distinct forms of (un)naturalness that are important for the CRISPR/Cas debate are what Siipi (2008) calls history-based and property-based (un)naturalness.

While history-based naturalness focuses on the origin of an entity or the kind of modification it has gone through, property-based naturalness looks at the current properties of an entity disregarding the way they came into existence.[6] For history-based naturalness, the human involvement is the point of reference to decide whether an entity is natural or not. Here, any human interference can itself be viewed as unnatural or it can gradually differ with regard to the intensity of interference. In the first case an entity can only be either natural or unnatural. In the second case different degrees of naturalness are possible. For property-based naturalness, Siipi (2008) identifies different comparative models that are possible to base a decision on. With regard to CRISPR/Cas, a 'historically natural entity', that is a wild plant with its molecular composition, is used as a point of reference.

Proponents of CRISPR/Cas as a new breeding technique mainly argue with naturalness in the property-based form. By indicating that the molecular structure of synthesized genes can be identical with natural genes and by arguing that rewilding leads to a back-to-nature crop (Palmgren et al., 2015), they interpret naturalness as similarity to historically natural entities. Wild plants are used as an ideal comparative model and therefore the new breeds with similar properties as the wild forms are viewed as natural. The more similarities exist, the more natural is the new entity. According to this line of reasoning, cisgenic plants are more natural than former GM plants because they remain within species boundaries. Furthermore, CRISPR/Cas genome editing can be viewed as natural because it is an action that is similar to a biological mechanism (Belhaj *et al.*, 2015). This is the second way of understanding the property-based naturalness with a historically natural entity as comparative model. Here the human action is compared with a naturally occurring process (Siipi, 2008). As mentioned earlier, CRISPR/Cas as an action applied by the plant breeder in the laboratory is based on and very similar to a mechanism naturally occurring in bacterial cells. Thus, from this perspective CRISPR/Cas is a kind of 'discovery' and as a new breeding technique it is only mimicking nature. Understanding new technologies as biomimicry of natural technologies is typical for what Zwart (2009) calls a techno-science perspective on nature, where nature is viewed as a great pool of products, processes and an inspiring source of knowledge, that has to be discovered, explored and imitated.[7] Nature provides the scientist with tools and techniques he or she can use to develop new products and procedures to solve human problems (Blok and Gremmen, 2016). This can be understood as collaboration with nature or as control over nature to optimize natural processes in light of human needs and finally 'to design the ideal plant type' (Koornneef and Stam, 2001). In both cases the categorization of a product or a process as natural serves as moral relief: according to this techno-science understanding of naturalness, no moral concerns arise as long as the new product could exist at least theoretically without human interference (Weigel, 2017). From this perspective the product is morally acceptable because it is natural in this specific sense.

In contrast to the techno-science perspective on breeding, opponents of CRISPR/Cas mainly focus on the process of human interference in nature. Here we find what Siipi (2008) calls the history-based naturalness concept. According to this understanding, CRISPR/Cas-modified plants are more unnatural than plants that are a result of traditional cross-breeding techniques because the former have undergone greater human-caused change processes than traditionally bred plants. This is even true if the genotype of the plants is similar. It is the processes

that give rise to moral concerns. The demand to regulate cisgenically modified plants under the GM regulations in the EU shows that the transgression of species boundaries, and thus the unnaturalness of the product, is not the key concern that is raised – rather, it is the process of human intrusion into the cell. This is part of a general criticism of technologies that infringe on natural processes. Opponents view these new plants as one step further to a biofact (Karafyllis, 2003), a neologism formed from 'bios' and 'artefact', where more and more key aspects of nature are lost. The term biofact expresses the concern that nature is no longer the non-human in the Aristotelean understanding – that which comes into being – but is shaped and designed by technology and becomes an artificially created entity. For this group the fundamental question still arises whether the acquisition of natural processes for human purpose is legitimate. Molecular biology allows what Lee (2003) calls an ontological transformation of living organisms from natural beings into 'biotic artefacts'. Opponents criticize that nature becomes from something that grows and develops to something that is made (Dabrock, 2009). For the opponents of CRISPR/Cas, this technology is not a step further to more naturalness by learning from nature, but a step further to achieve control of nature and its processes and therefore a highly moral question.

What becomes apparent by specifying the term 'natural' or 'naturalness' is that the different levels of acceptance of the new technology are not the result of knowledge deficits by consumers or citizens, but mainly caused by different values about the way humans should be allowed to interfere with nature. The diagnostic perspective on naturalness helps understand why the societal debate on the general acceptance of genome editing continues despite the shift from transgenesis to cisgenesis. The idea that 'nature-identity' or molecular equivalence is the panacea for enabling genetic modification of plants without ethical discourse is an illusion. CRISPR/Cas genome editing is still not viewed as ethically neutral. As is typical for value-plural societies, there is no common perspective on how far humans should be allowed to infringe on natural processes. If the basic attitude towards nature orients itself towards the product, the ethical debate will be less intensive than the former GM debate (Delwaide *et al.*, 2015; Edenbrandt *et al.*, 2018). If, however, a process perspective dominates, this could further put in question the commonality of values in science and society and thus intensify ethical tensions. As mentioned earlier, governance plays a key role in balancing different values and interests within society. The minimum governments can do in such a situation of diverging public perceptions is to guarantee the freedom of choice between different products.

Governing CRISPR/Cas: Conflicting Rights and Interests

Regulation, ban or moratorium, and laissez-faire are three different general approaches to govern a new technology. Due to the fast spread of CRISPR/Cas, together with current international trade agreements as well as heterogeneous risk perception on the technology, a global ban or a moratorium do not seem to be viable options. Thus, for the long term at least, the decision on how the use of CRISPR/Cas-generated crops in agriculture should be governed has to be taken between regulation and laissez-faire. While in a sense this can be treated as a technical and 'pragmatic' issue, it too has important ethical repercussions. In fact, this ethical dimension of governance has been overlooked in many recent publications on the topic (Huang *et al.*, 2016; Ishii and Araki, 2016; Pollock, 2016), whose authors effectively advocated for discarding supposedly 'irrational fears' of the wider public and a product-based regulation of CRISPR/Cas products – which, in the case of nGMOs, would effectively mean non-regulation or, more precisely, regulation in the same limited extent as for conventionally bred crops. This amounts to circumventing the societal and value debates discussed in the previous two sections. However, one of the main goals of governance is to achieve a compromise in

such debates. Governance is a balancing act between the plural values present in a society, and this balancing act is required if a governance solution is to be legitimate (Vatn, 2016). In the case of GM governance, this is particularly true because of the ethical imperative to include stakeholders in decisions that involve their deeply held values (Jasanoff *et al.*, 2015). A more pragmatic argument is that ignoring consumer preferences might render GM regulation irrelevant, leading to implicit or explicit boycotts of GM products: 'the key issue is not whether new crop varieties are as safe as those developed by conventional plant breeding and thus fall outside the scope of current GMO legislation, but whether society perceives them as such' (Malyska *et al.*, 2016, p. 532; see also Kuzma, 2016).

This being said, there are essentially two reasons why CRISPR/Cas applications should be regulated. The first reason is consumer sovereignty, the right for consumers to reject some products (e.g. a CRISPR/Cas-generated nGMO) for whatever (normative) reasons they have. The second reason is that, despite its relative precision, CRISPR/Cas involves uncertain hazards – most notably, off-target modifications (Schaefer *et al.*, 2017), which also necessitate careful governance (Stirling, 2010; Kvakkestad and Vatn, 2011). Because governance is a means to strike a balance between conflicting rights, values and interests, its potentials and limitations are also ethically relevant.

In the following, we would like to discuss two issues pertaining to the governance of CRISPR/Cas: first, the consequences of non-traceability of genetic modifications in the resultant nature-identical GM organism; and second, the relevance of potentially wide access to the CRISPR/Cas technique due to its low-cost nature and relative simplicity.

Non-traceability of modifications

The supposed impossibility of determining in a nature-identical product of CRISPR/Cas genome editing how it came into being precludes effective monitoring, which makes product-based regulation technically impractical – it would effectively amount to non-regulation. Current GM regulations largely depend on the existence of an exogenous DNA sequence in the resultant organism, which is not present in nature-identical CRISPR/Cas products. While it has been argued that CRISPR/Cas leaves small amounts of foreign DNA in the genome (Kim and Kim, 2016), conventional analytical methods such as PCR (polymerase chain reaction) cannot detect such evidence. Non-traceability of modifications calls for process-based regulatory approaches. A product-based approach would effectively amount to laissez-faire, which would violate the principle of consumer sovereignty, given that many people oppose GM crops irrespective of whether they are molecularly equivalent to 'natural' varieties (nGMOs) or not.

However, in contrast to product-based regulation as adopted in the USA or Canada, process-based regulation is much more difficult to set up, especially given the more diversified use of the technology encompassing many more actors in the future (see below). Here, the EU has the advantage of having experience in process-based governance. However, at the time of writing, the European Food Safety Authority has not issued a concluding opinion whether CRISPR/Cas products should generally be considered GMOs, and whether current regulations apply. From a 'technical' governance point of view, it is not relevant whether nature-identical CRISPR/Cas products are considered GMOs or not because non-traceability effectively precludes product-based regulation. It is mainly economic (because regulation imposes costs) and ethical reasons that necessitate such categorizations, as explained above. Calls in the CRISPR/Cas literature to introduce a product-based governance system for foods, which would effectively lead to nGMOs being indistinguishable from conventionally bred products (e.g. Huang et al., 2016; Ishii and Araki, 2016; Pollock, 2016), effectively ignore the many ethical tensions and challenges posed by CRISPR/Cas products, which go well beyond risk assessment issues.

There is a clear discrepancy between consumers' right to know (as advocated, for instance, by the proponents of GM labeling) and, in a wider sense, consumer sovereignty on the one hand, and the possibility to trace CRISPR/Cas-induced modifications in the end product on the other hand. While it can be argued that CRISPR/Cas may lower the reluctance of consumers towards GMOs (Bartkowski et al., 2017), many consumers will still reject CRISPR/Cas on the grounds of it being 'unnatural'. Paradoxically, CRISPR/Cas could even increase the divide between those opposing GM foods and those embracing the promises of biotechnology. If it would contribute to reducing the opposition in some parts of society and possibly even lead to admission of cisgenesis in organic agriculture (Ronald and Adamchak, 2008), those who still view cisgenesis as 'unnatural' might entrench even more, as they would likely feel that their choices are being further circumscribed. This suggests that, to be legitimate, CRISPR/Cas governance needs a broad societal debate (Jasanoff et al., 2015). As shown above, to date this debate has been taking place mainly in the context of organic agriculture, where the divides are perhaps deepest. With increased success and popularity of CRISPR/Cas, it can be expected to spread to the wider society. The question is, though, whether this will happen soon enough before path dependencies arise and irreversible decisions are made.

Decentralization of knowledge and use

Path dependencies can arise, among other things, if CRISPR/Cas leads to a diversification of the green biotechnology market. Being relatively cheap and easy-to-use, as compared with other GM technologies, CRISPR/Cas genome editing has the potential to diversify the market for biotechnology by reducing market entry barriers. Before the advent of CRISPR/Cas, genetic engineering was an enterprise with very high upfront investment costs (including the costs of know-how and laboratory equipment, as well as high opportunity costs of time invested in each innovation), which partly explains the highly concentrated market structure of biotech industries. Now, the technique can arguably be applied at very low cost also by small biotech firms, non-profit organizations or public institutions (see Ledford, 2015). Of course, because its application requires profound knowledge about genetics and biotechnology, it is unlikely that it could be used even by individual farmers in the near future. But it is definitely not necessary anymore to have the capital of multinational biotech companies to engage in genetic engineering.

A consequence of this could be a significant change in the market structure of biotech industries. As of now, this is a highly concentrated market (Howard, 2015), as highlighted by the recent merger of Bayer and Monsanto. The market diversification potential of CRISPR/Cas genome editing might lead to a deconsolidation of the market, since big players would lose some of their comparative advantage in terms of scale effects and capital availability. As shown by Brinegar *et al.* (2017), the number of patent applications filed for CRISPR/Cas applications has increased tremendously since 2012 (see also Ledford, 2015). Furthermore, despite ongoing legal disputes (Contreras and Sherkow, 2017a, 2017b; Horn, 2017), the most basic CRISPR patents are all possessed by firms linked to public institutions in the USA, which have largely committed to issuing non-exclusive licenses for commercial use of the technology (McGuire, 2016). Thus, while the CRISPR/Cas tools are not freely available, the obstacles to their use are significantly lower than is the case with GM technologies developed by commercial entities.

While generally positive, the widespread use of CRISPR/Cas, including by non-profit organizations and smaller firms, might lead to an increase in biotech-related risks, both because of the diversification of sources of genetic modifications and because of the relative inexperience of some of the actors involved. Even with CRISPR/Cas, genetic engineering is a demanding and complex enterprise, involving large risks and high levels of uncertainty (Kvakkestad and Vatn, 2011). The current biotech market players have many years of experience in dealing with such risks and uncertainty. In contrast, the limited ability of smaller players to handle even existing hazards poses new governance challenges.

An ethical tension arises between the legitimate hope of deconsolidation and demonopolization of the biotech market that might be spurred by CRISPR/Cas genome editing, and the risks this could create in terms of control difficulties, increase in bureaucracy (to handle the newly created diversity) and even environmental and health risks if the diversification leads to less stringent controls. Furthermore, it is an open question how the intellectual property rights will develop and how freely the new technology can be applied in the future (Sherkow, 2016). This means that the advent of CRISPR/Cas may have repercussions for the critical debates on food sovereignty (Kloppenburg, 1988, 2014; Luby *et al.*, 2015) and 'patenting life' (Hettinger, 1995; Sherkow and Greely, 2015). These issues are, however, beyond the scope of this chapter.

Conclusions

The question how CRISPR/Cas should be regulated clearly goes beyond pure scientific knowledge about the new technology. Rather, it includes fundamental moral judgments about the need for and the way of adapting nature for human needs. In plant breeding this is a particularly challenging question, as it is the ultimate goal of breeding to adjust plants for human purposes. If we categorize the overall moral debate into two general groups of arguments, teleological and principle-based, we can observe a shift in the focus of the discussion when comparing the former GM debate and the new debate on CRISPR/Cas. In the former GM debate, ethical tensions arose *within* the teleologically inspired concerns about risks and benefits of the technology as well as *between* teleological and principle-based concerns. In contrast, we now observe that ethical tensions on CRISPR/Cas mainly arise *within* the principle-based perspective. Naturalness as a principle and a moral category no longer provides a clear argument for rejecting the new technology, as was the case in the former GM debate. Rather, it can be invoked by the technology's proponents as well, due to the supposed nature-identity of CRISPR/Cas-modified products. Therefore, ethical tensions now arise between different concepts of naturalness; they have moved away from the former focus on risk but also from the general tension between teleological and principle ethics. Whether this new line of discussion will deepen or blur the division between proponents and opponents of the new technology will strongly depend on whether natural sciences start regarding the ethical dimension intrinsic rather than extrinsic to their task (Bruce, 2002).

This shift in the debate has repercussions for governance in that it suggests that the often called-for 'science-based' regulation, based on the idea of nature-identity, would be misguided by ignoring the new ethical tensions. Furthermore, some technical characteristics of the technology, like the non-traceability of modifications, also determine how CRISPR/Cas-modified plants can be governed. The dynamics of technological innovation put the development of adjusted and flexible regulatory governance structures under a time constraint. The technological development threatens to overrun the required discursive processes of collective decision making. To avoid reducing the role of governance to *ex-post* legitimation, there is an urgent need for a new institutional set up to establish a continuous discourse between science and society. Only then can the ethical tensions created by CRISPR/Cas be navigated and solved. They can be viewed as part of a general unease, skepticism and distrust towards private and public actors and the decision-making processes along the entire food chain that result from recent developments in plant breeding in particular and agricultural production in general, fueling, among others, a countermovement under the umbrella of food sovereignty. The underlying ethical tensions must be addressed by means of a broad societal discourse and involvement of the general public in decision making. This would also contribute to the legitimacy of the governance structure decided on, informed by the discourse.

Notes

[1] CRISPR stands for Clustered Regularly Interspaced Short Palindromic Repeats and is a specific type of DNA sequence; Cas is a CRISPR-associated protein.

[2] However, the possibility of completely non-traceable modification has been questioned (Kim and Kim, 2016). We return to this issue later.

[3] For more details about the history of the relevant discoveries and also about the controversial attribution of scientific credits, we refer the reader to Lander (2016) and Ledford (2016).

[4] At the time of writing, the EU has been awaiting the European Court of Justice' decision about whether non-transgenic genome editing should be considered as (targeted) mutagenesis and thus exempted from GM regulation.

[5] 'Organic Agriculture should sustain and enhance the health of soil, plant, animal, human and planet as one and indivisible. This principle points out that the health of the individuals and communities cannot be separated from the health of ecosystems – healthy soils produce healthy crops that foster the health of animals and people. Health is the wholeness and integrity of living systems. It is not simply the absence of illness, but the maintenance of physical, mental, social and ecological well-being. Immunity, resilience and regeneration are key characteristics of health' (IFOAM, 2014, p. 9).

[6] Note the obvious correspondence with the distinction between process-based and product-based regulation of GMOs (see section on governance).

[7] This perspective is typical of economics, where biodiversity is viewed as a 'portfolio' of (potential) assets and thus ascribed option value (Bartkowski, 2017).

References

Andersen, M.M. *et al.* (2015) Feasibility of new breeding techniques for organic farming. *Trends in Plant Science* 20, 426–434. doi:10.1016/j.tplants.2015.04.011

Araki, M. and Ishii, T. (2015) Towards social acceptance of plant breeding by genome editing. *Trends in Plant Science* 20, 145–149. doi:10.1016/j.tplants.2015.01.010

Baker, M. (2014) Gene editing at CRISPR speed. *Nature Biotechnology* 32, 309–312. doi:10.1038/nbt.2863

Baltes, N.J., Gil-Humanes, J. and Voytas, D.F. (2017) Genome engineering and agriculture: Opportunities and challenges. *Progress in Molecular Biology and Translational Science* 149, 1–26. doi:10.1016/bs.pmbts.2017.03.011

Bartkowski, B. (2017) Are diverse ecosystems more valuable? Economic value of biodiversity as result of uncertainty and spatial interactions in ecosystem service provision. *Ecosystem Services* 24, 50–57. doi:10.1016/j.ecoser.2017.02.023

Bartkowski, B., Theesfeld, I., Pirscher, F. and Timaeus, J. (2017) Snipping around for food: Economic, ethical and policy implications of CRISPR/Cas genome editing. Working paper.

Belhaj, K. *et al.* (2015) Editing plant genomes with CRISPR/Cas9. *Current Opinion in Biotechnology* 32, 76–84. doi:10.1016/j.copbio.2014.11.007

Bio Suisse (2016) CRISPR/CAS und die Biobranche. Available at: https://www.bio-suisse.ch/media/Ueberuns/Medien/unsere_meinung_zu_crispr_cas.pdf (accessed 18 July 2017).

Bioland (2016) Gentechnikverfahren CRISPR/Cas ist absolutes No-Go für Biolandwirtschaft. Available at: http://www.bioland.de/presse/presse-detail/article/bioland-leben-ist-nicht-programmierbar-gentechnikverfahren-crisprcas-ist-absolutes-no-go-fuer-biolandwirtschaft.html (accessed 18 July 2017).

Blok, V. and Gremmen, B. (2016) Ecological innovation: Biomimicry as a new way of thinking and acting ecologically. *Journal of Agricultural and Environmental Ethics* 29, 203–217. doi:10.1007/s10806-015-9596-1

Brinegar, K. *et al.* (2017) The commercialization of genome-editing technologies. *Critical Reviews in Biotechnology* 37, 924–932. doi:10.1080/07388551.2016.1271768

Brooks, C., Nekrasov, V., Lippman, Z.B. and Eck, J.V. (2014) Efficient gene editing in tomato in the first generation using the Clustered Regularly Interspaced Short Palindromic Repeats/CRISPR-Associated9 System. *Plant Physiology* 166, 1292–1297. doi:10.1104/pp.114.247577

Bruce, D.M. (2002) A social contract for biotechnology: Shared visions for risky technologies? *Journal of Agricultural and Environmental Ethics* 15, 279–289. doi:10.1023/A:1015738727342

Cardi, T. (2016) Cisgenesis and genome editing: Combining concepts and efforts for a smarter use of genetic resources in crop breeding. *Plant Breeding* 135, 139–147. doi:10.1111/pbr.12345

Cong, L. *et al.* (2013) Multiplex genome engineering using CRISPR/Cas systems. *Science* 339, 819–823. doi:10.1126/science.1231143

Contreras, J.L. and Sherkow, J.S. (2017a) CRISPR, surrogate licensing, and scientific discovery. *Science* 355, 698–700. doi:10.1126/science.aal4222

Contreras, J.L. and Sherkow, J.S. (2017b) Patent pools for CRISPR technology – Response. *Science* 355, 1274–1275. doi:10.1126/science.aan0818

Dabrock, P. (2009) Playing God? Synthetic biology as a theological and ethical challenge. *Systems and Synthetic Biology* 3, 47. doi:10.1007/s11693-009-9028-5

Delwaide, A.-C. *et al.* (2015) Revisiting GMOs: Are there differences in European consumers' acceptance and valuation for cisgenically vs transgenically bred rice? *PLOS One* 10, e0126060. doi:10.1371/journal.pone.0126060

Edenbrandt, A.K., Gamborg, C. and Thorsen, B.J. (2018) Consumers' preferences for bread: Transgenic, cisgenic, organic or pesticide-free? *Journal of Agricultural Economics* 69, 121–141. doi:10.1111/1477-9552.12225

Gaskell, G. *et al.* (2010) *Europeans and Biotechnology in 2010: Winds of Change?* A report to the European Commission's Directorate-General for Research. European Commission, Brussels.

Hettinger, N. (1995) Patenting life: Biotechnology, intellectual property, and environmental ethics. *Boston College Environmental Affairs Law Review* 22, 267–305.

Horn, L. (2017) Patent pools for CRISPR technology. *Science* 355, 1274–1274. doi:10.1126/science.aan0515

Howard, P.H. (2015) Intellectual property and consolidation in the seed industry. *Crop Science* 55, 2489. doi:10.2135/cropsci2014.09.0669

Huang, S., Weigel, D., Beachy, R.N. and Li, J. (2016) A proposed regulatory framework for genome-edited crops. *Nature Genetics* 48, 109–111. doi:10.1038/ng.3484

IFOAM (2014) The IFOAM norms for organic production and processing. Available at: https://www.ifoam.bio/sites/default/files/ifoam_norms_version_july_2014.pdf (accessed 1 November 2017).

Ishii, T. and Araki, M. (2016) Consumer acceptance of food crops developed by genome editing.

Plant Cell Reports 35, 1507–1518. doi:10.1007/s00299-016-1974-2

Jacobs, T.B., LaFayette, P.R., Schmitz, R.J. and Parrott, W.A. (2015) Targeted genome modifications in soybean with CRISPR/Cas9. *BMC Biotechnology* 15, 16. doi:10.1186/s12896-015-0131-2

Jasanoff, S., Hurlbut, J.B. and Saha, K. (2015) CRISPR democracy: Gene editing and the need for inclusive deliberation. *Issues in Science and Technology* 32, 37–49.

Jinek, M. *et al.* (2012) A programmable dual-RNA-guided DNA endonuclease in adaptive bacterial immunity. *Science* 337, 816–821. doi:10.1126/science.1225829

Karafyllis, N.C. (2003) Das wesen der biofakte. In: Karafyllis, N.C. (ed.) *Biofakte – Versuch Über Menschen Zwischen Artefakt Und Lebewesen*. Mentis, Paderborn, Germany, pp. 11–26.

Khlestkina, E.K. and Shumny, V.K. (2016) Prospects for application of breakthrough technologies in breeding: The CRISPR/Cas9 system for plant genome editing. *Russian Journal of Genetics* 52, 676–687. doi:10.1134/S102279541607005X

Kim, J. and Kim, J.-S. (2016) Bypassing GMO regulations with CRISPR gene editing. *Nature Biotechnology* 34, 1014–1015. doi:10.1038/nbt.3680

Kloppenburg, J.R. (1988) *First the Seed: The Political Economy of Plant Biotechnology, 1492–2000*. Cambridge University Press, Cambridge, UK.

Kloppenburg, J. (2014) Re-purposing the master's tools: The open source seed initiative and the struggle for seed sovereignty. *Journal of Peasant Studies* 41, 1225–1246. doi:10.1080/03066150.2013.875897

Koornneef, M. and Stam, P. (2001) Changing paradigms in plant breeding. *Plant Physiology* 125, 156–159. doi:10.1104/pp.125.1.156

Kuzma, J. (2016) Reboot the debate on genetic engineering. *Nature* 531, 165–167. doi:10.1038/531165a

Kvakkestad, V. and Vatn, A. (2011) Governing uncertain and unknown effects of genetically modified crops. *Ecological Economics* 70, 524–532. doi:10.1016/j.ecolecon.2010.10.003

KWS (2015) Präzise Verfahren. Naturidentische Ergebnisse. KWS im Dialog: Aktuelles für Entscheidungsträger. Available at: https://biokulturorg.files.wordpress.com/2017/01/kws_in_dialog-kopie.pdf (accessed 1 November 2017).

Lammerts van Bueren, E.T., Struik, P.C., Tiemens-Hulscher, M. and Jacobsen, E. (2003) Concepts of intrinsic value and integrity of plants in organic plant breeding and propagation. *Crop Science* 43, 1922–1929. doi:10.2135/cropsci2003.1922

Lander, E.S. (2016) The heroes of CRISPR. *Cell* 164, 18–28. doi:10.1016/j.cell.2015.12.041

Ledford, H. (2015) CRISPR, the disruptor. *Nature New* 522, 20. doi:10.1038/522020a

Ledford, H. (2016) The unsung heroes of CRISPR. *Nature* 535, 342–344. doi:10.1038/535342a

Lee, K. (2003) *Philosophy and Revolutions in Genetics: Deep Science and Deep Technology, Renewing Philosophy*. Palgrave Macmillan, New York.

Li, Z. *et al.* (2015) Cas9-Guide RNA directed genome editing in soybean. *Plant Physiology* 169, 960–970. doi:10.1104/pp.15.00783

Luby, C.H., Kloppenburg, J., Michaels, T.E. and Goldman, I.L. (2015) Enhancing freedom to operate for plant breeders and farmers through open source plant breeding. *Crop Science* 55, 2481. doi:10.2135/cropsci2014.10.0708

Malyska, A., Bolla, R. and Twardowski, T. (2016) The role of public opinion in shaping trajectories of agricultural biotechnology. *Trends in Biotechnology* 34, 530–534. doi:10.1016/j.tibtech.2016.03.005

McGuire, L. (2016) Genome editing: Broad Institute keeps CRISPR tools open. *Nature* 534, 37–37. doi:10.1038/534037a

Meinard, Y. (2017) What is a legitimate conservation policy? *Biological Conservation* 213, Part A, 115–123. doi:10.1016/j.biocon.2017.06.042

Nuijten, E., Messmer, M.M. and Lammerts van Bueren, E.T. (2016) Concepts and strategies of organic plant breeding in light of novel breeding techniques. *Sustainability* 9, 18. doi:10.3390/su9010018

Palmgren, M.G. *et al.* (2015) Are we ready for back-to-nature crop breeding? *Trends in Plant Science* 20, 155–164. doi:10.1016/j.tplants.2014.11.003

Pollock, C.J. (2016) How should risk-based regulation reflect current public opinion? *Trends in Biotechnology* 34, 604–605. doi:10.1016/j.tibtech.2016.05.002

Ried, J., Braun, M. and Dabrock, P. (2011) Unbehagen und kulturelles Gedächtnis. Beobachtungen zur gesellschaftlichen Deutungsunsicherheit gegenüber Synthetischer Biologie. In: Dabrock, P., Bölker, M., Braun, M. and Ried, J. (eds) *Was Ist Leben – Im Zeitalter Seiner Technischen Machbarkeit? Beiträge Zur Ethik Der Synthetischen Biologie*. Verlag Karl Alber, Freiburg, Germany, pp. 345–368.

Ronald, P.C. and Adamchak, R.W. (2008) *Tomorrow's Table: Organic Farming, Genetics, and the Future of Food*. Oxford University Press, New York.

Schaefer, K.A. *et al.* (2017) Unexpected mutations after CRISPR-Cas9 editing in vivo. *Nature Methods* 14, 547–548. doi:10.1038/nmeth.4293

Schouten, H.J., Krens, F.A. and Jacobsen, E. (2006) Cisgenic plants are similar to traditionally bred plants: International regulations for genetically modified organisms should be altered to exempt cisgenesis. *EMBO Reports* 7, 750–753. doi:10.1038/sj.embor.7400769

Sherkow, J.S. (2016) CRISPR: Pursuit of profit poisons collaboration. *Nature News* 532, 172. doi:10.1038/532172a

Sherkow, J.S. and Greely, H.T. (2015) The history of patenting genetic material. *Annual Reviews of Genetics* 49, 161–182. doi:10.1146/annurev-genet-112414-054731

Siipi, H. (2008) Dimensions of naturalness. *Ethics & the Environment* 13, 71–103.

Steinbrecher, R.A. (2015) Genetic engineering in plants and the 'New Breeding Techniques (NBTs)'. Inherent risks and the need to regulate. Econexus Briefing. Available at: http://nuffield bioethics.org/wp-content/uploads/genome-editing-evidence-EcoNexus.pdf (accessed 1 November 2017).

Stirling, A. (2010) Keep it complex. *Nature* 468, 1029–1031. doi:10.1038/4681029a

Svitashev, S. *et al.* (2015) Targeted mutagenesis, precise gene editing, and site-specific gene insertion in maize using Cas9 and guide RNA. *Plant Physiology* 169, 931–945. doi:10.1104/pp.15.00793

taz (2016) Ökoforscher über neue Gentech-Methode: 'CRISPR hat großes Potenzial'. *Tageszeitung*. Available at: http://www.taz.de/!5290509/ (accessed 1 November 2017).

Then, C. (2016) *Synthetic Gene Technologies Applied in Plants and Animals Used for Food Production: Overview on Patent Applications on New Techniques for Genetic Engineering and Risks Associated with These Methods.* Testbiotech, Munich, Germany.

Ueta, R. *et al.* (2017) Rapid breeding of parthenocarpic tomato plants using CRISPR/Cas9. *Scientific Reports* 7. doi:10.1038/s41598-017-00501-4

van Haperen, P.F., Gremmen, B. and Jacobs, J. (2012) Reconstruction of the ethical debate on naturalness in discussions about plant-biotechnology. *Journal of Agricultural and Environmental Ethics* 25, 797–812. doi:10.1007/s10806-011-9359-6

Vatn, A. (2016) *Environmental Governance: Institutions, Policies and Actions.* Edward Elgar, Cheltenham, UK.

Waltz, E. (2016) Gene-edited CRISPR mushroom escapes US regulation. *Nature* 532, 293. doi:10.1038/nature.2016.19754

Weigel, D. (2017) Ethische fragen spielen keine rolle – einsatz chancen und risiken von CRISPR/Cas bei pflanzen. *Forschung und Lehre* 1, 22–23.

Woo, J.W. *et al.* (2015) DNA-free genome editing in plants with preassembled CRISPR-Cas9 ribonucleoproteins. *Nature Biotechnology* 33, 1162–1164. doi:10.1038/nbt.3389

Zwart, H. (2009) Biotechnology and naturalness in the genomics era: Plotting a timetable for the biotechnology debate. *Journal of Agricultural and Environmental Ethics* 22, 505–529. doi:10.1007/s10806-009-9178-1

11

New Technology, Cognitive Bias and Ethical Tensions in Entrepreneurial Commercialization: The Case of CRISPR

Desmond Ng[1]* and Harvey S. James, Jr.[2]

[1]Department of Agricultural Economics, Texas A&M University, College Station, Texas, USA; [2]Division of Applied Social Sciences, University of Missouri, Columbia, Missouri, USA

Introduction

The identification and exploitation of external opportunities are widely recognized as central functions of entrepreneurship. Identification refers to the creation of new inventions or technologies and exploitation refers to efforts to bring them to market. However, both the development of technologies and efforts to commercialize them depend on the resources and expertise of others. For instance, a study of randomly selected drugs from ten pharmaceutical firms revealed the average cost of bringing a new drug through the approval process to market was nearly US$2.6 billion (2013 dollars) (DiMasi *et al.*, 2016), suggesting that entrepreneurial R&D start-ups will need to depend on the development, manufacturing, marketing and/or distribution assets of commercializing partners (Ng, 2011).

Because the commercialization of an entrepreneur's inventions depends on the assets and resources of its various stakeholders, an entrepreneur has an ethical responsibility to the welfare of these stakeholders. However, the entrepreneur also has a responsibility to act in accordance with the demands of society (Joyner and Payne, 2002; Noland

and Phillips, 2010). This is because breakthrough innovations can have widespread and long-term effects on societal members (e.g. Edison's light bulb, Ford's assembly line production, the internet) that extend beyond those stakeholders who have contributed to an entrepreneur's commercializing efforts. Thus, entrepreneurs developing and seeking to commercialize new technologies face a fundamental ethical tension between the interests of their stakeholders and the interests of society.

However, the ability of entrepreneurial inventors to navigate this ethical tension is complicated by the fact that they operate in highly complex and novel decisions settings and that entrepreneurs can fail to account for stakeholders that are affected by their decision making in these environments (James *et al.*, 2016). Thus, entrepreneurs not only face an ethical tension between the interests of stakeholders and the interests of society, but also can lapse in their ethical obligations to their stakeholders. Research on entrepreneurial cognition has shown that complexity and novelty can result in 'egocentric' and 'myopic' biases that impair an entrepreneur's ethical judgments with respect to broader societal groups (e.g. Lepoutre and

---

* Corresponding author. E-mail: dng@tamu.edu

Heene, 2006; Dew and Sarasvathy, 2007). In examining such arguments, James *et al.* (2016) used the World Values Survey to show a connection between complexity and novelty and poor ethical judgments that is consistent with the onset of confirmation biases (Palich and Bagby, 1995; Nickerson, 1998) and inward biases (Kahneman and Lovallo, 1993; Kahtri and Ng, 2000). While James *et al.* (2016) offer insights to understanding biases surrounding an entrepreneur's ethical judgments, they did not develop the argument with an entrepreneur's commercialization of new technology in mind. Yet, commercialization of breakthrough innovations not only are often complex and novel, but also are capable of having wide and long-term social and moral consequences that are not considered by the creators of those innovations.

For instance, recent investor attention in the CRISPR (clustered regularly interspaced short palindromic repeats) genome-editing tool has sparked widespread interest in its commercial potential. However, the commercialization of CRISPR technology involves a complexity of stakeholder resources and expertise that enable entrepreneurial firms to discover its various applied uses. Furthermore, as CRISPR reflects a genuinely novel breakthrough in genomic research, such complexity and novelty might impair an entrepreneur's ability to anticipate the ethical consequences of these commercial discoveries, as suggested by the James *et al.* (2016) thesis.

In this chapter, we reexamine James *et al.*'s (2016) concepts of complexity and novelty and their associated cognitive biases within the context of the commercialization of CRISPR technology. We highlight propositions to explain an entrepreneur's ethics in complex and novel decision settings and then use those insights to comment on the ethical tensions and challenges associated with the commercialization of the CRISPR technology.

Relevant Details About the CRISPR Technology

CRISPR has been recognized as one of the most significant breakthroughs in genomic science. CRISPR offers a revolutionary genomic editing tool to target, delete and replace specific genes in any living organism and has a wide number of commercial applications, including but not limited to curing genetic defects and offering various agronomic improvements in agricultural productivity (Lewis, 2015).

There are variations of CRISPR processes, but the one that has generated the most promising results derives from the work of scientists Jennifer Doudna and Feng Zhang and their colleagues. They made possible the development of a 'guide RNA', which allows scientists to make a genetic sequence or guideRNA to match the piece of DNA that they want to modify. The guideRNA is then added to the CAS9 protein, which searches out cells containing the genetic sequences that matches the guideRNA. The guideRNA cuts out the portion of the genetic sequence to be replaced with another desirable gene sequence. By using this search and replace function of the CRISPR/Cas9 system, genetic editing not only offers an unprecedented level of control in editing specific portions of an organism's DNA, but also has broad applications in the development and commercialization of a wide range of therapeutic and agronomic products (Lewis, 2015; Booth, 2016).

Most notably, CRISPR/Cas9 offers a potentially wide range of applications to the medical and drug industry that involve reducing the cost and time of drug development, the development of better animal models for the treatment of human diseases, and the replacement of human organs with those from animal sources (Lewis, 2015; McKinsey & Company, 2017). Furthermore, there are also important commercial applications to the agriculture industry, where it can be used to engineer microbes that produce biofuels more efficiently, modify genes in plant seeds with a greater tolerance to herbicide, improve drought tolerance (McKinsey & Company, 2017) and speed up the development of new crops (Markman, 2017). For instance, scientists at Cold Harbor Laboratory in New York showed how CRISPR/Cas9 can be used to modify the genes responsible for the production of excessive branches in the tomato plant. By editing the genes responsible for these excessive branches, the removal of excessive branches reduced the cost in harvesting tomatoes while increasing tomato yields. This genomic editing also allowed the tomato to stay on the plant longer, which reduced incidence of bruising damage to the fruit (Markman, 2017). Similarly, research scientists at Penn State used the CRISPR/Cas9 system to remove the gene that causes browning in mushrooms (Markman, 2017). CRISPR/Cas9 also offers the potential to resolve some of the more fundamental challenges facing the agrochemical industry. One of the long-term challenges facing this industry is that resistance to synthetic pesticide has been steadily growing over the past several decades, but the rate of discovery of new pesticides has 'gone almost to zero in the last 10 years or so' (Borel, 2017, n.p.). Some have suggested that the CRISPR/Cas9 technology offers significant opportunities to revitalize the development of new bio-based pesticides in reversing this declining trend (Borel, 2017). Lastly, business commentators have also claimed that the CRISPR/Cas9 system may even offer the potential to solve world hunger (Markman, 2017).

Conceptual Framework

In response to exploiting the opportunities made available by the CRISPR/Cas9 system, we define the entrepreneur by an individual's ability to identify and exploit valued opportunities (Klein, 2008). In the context of this chapter, this involves identifying and exploiting the commercialization opportunities of the CRISPR genomic editing technology. This functional definition of the entrepreneur suggests that an entrepreneur's ability to commercialize the CRISPR technology is based not only on the inventive efforts of the individual, but also on those stakeholders who can provide the skills, expertise, investment capital and technology that is necessary in bringing the technology to market. That is, the commercialization of CRISPR involves drawing on the resources of its varied stakeholders as a means to enact

an entrepreneur's valued ends (e.g. Sarasvathy, 2001; see also Ng, 2015).

Stakeholders are defined as 'any group or individual who can affect or is affected by the achievement of an organization's objectives', and an entrepreneur's ethics is defined by their ability to be held accountable to stakeholders who 'affect or are affected' by their commercializing decisions (Freeman, 1984, p. 46).

An entrepreneur is accountable to two types of stakeholder groups: 'direct' and 'indirect' stakeholders. Direct stakeholders affect an entrepreneur's business objectives of profitability, growth and survival through the provision of resources and the development of ongoing relationships. Direct stakeholders include employees, strategic partners and stockholders in providing key resources in advancing the commercial potential of new technology. In contrast, indirect stakeholders are those who do not have a direct stake or investment in realizing the commercial potential of an entrepreneur's business but nonetheless contribute to and are affected by the firm's actions. For instance, while government agencies, such as the USDA, EPA and FDA, do not have a direct investment in an entrepreneur's business, they provide oversight and assurance to the public about the safety and benefits surrounding the regulation of new food and drug products.

As the commercialization of the CRISPR/Cas9 technology can influence both direct and indirect stakeholders, the developers of the technology and entrepreneurial CRISPR start-up firms (e.g. Intellia Therapeutics, Editas Medicine) are ethically accountable to these groups of stakeholders. Yet, as we will explain, an entrepreneur's ethical obligations operate in situations of complexity and novelty that can inhibit the entrepreneur from fully comprehending the consequences of their actions on others. Specifically, James *et al.* (2016) show that complex and novel decision settings render entrepreneurs vulnerable to confirmation and inward biases, which in turn affect ethical judgments favoring direct over indirect stakeholders (see Fig. 11.1).

In extending the insights of James *et al.*'s (2016) model, we claim that the

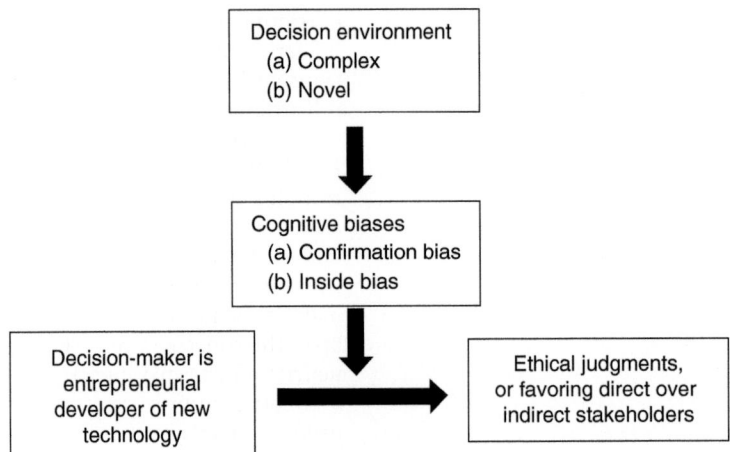

Fig. 11.1. Conceptual representation of entrepreneurial innovator's ethical judgment affected by complex and novel decision-making environments. (Adapted from Figure 1 in James *et al.*, 2016.)

commercialization of the CRISPR/Cas9 technology introduces a complexity and novelty where resulting biases prioritize the commercial interests of direct stakeholders and thus produce an ethical blindness in key decision makers. This is consistent with ethical blindness research that finds entrepreneurs can fail to do the right thing because their actions to satisfy the practical concerns of one group (i.e. direct stakeholders) can blind them from fulfilling their moral obligations to a broader indirect social group (Palazzo *et al.*, 2012; see also Chugh and Bazerman, 2007).

Complexity

Something is complex when there are many interrelated and interconnected parts that are difficult to understand and to assemble or disassemble without specialized knowledge. The CRISPR technology is difficult to understand and utilize without specialized scientific knowledge, thus making the technology complex. But the means to bring the technology to market also requires a complexity of direct and indirect stakeholders. Direct stakeholders include strategic partnerships and venture capitalists who provide advice and resources to overcome the specific operational challenges in the

commercialization of the CRISPR/Cas9 technology. For instance, the CRISPR start-up firm Intellia Therapeutics formed a commercializing partnership with Caribou Biosciences and Novartis to develop research and human therapeutic applications of CRISPR/Cas9 (Booth, 2016).

There are also important indirect stakeholders. Given the history of GM adoption, indirect stakeholders involving present and future consumers of CRISPR/Cas9 food- and drug-related products who can influence the commercial potential of this technology. Additionally, since there remains much uncertainty surrounding the safety of CRISPR/Cas9 products, government regulatory agencies, such as the USDA, EPA and FDA, will play a critical role in resolving this uncertainty and thus indirectly influence the commercial adoption of CRISPR/Cas9. Lastly, acceptance by other indirect stakeholders involving non-users of the CRISPR/Cas9-based products, such as affected agricultural producer groups, environmental, animal and other social activist groups, can influence its social acceptance and legitimacy.

Confirmation bias

The complexity of the CRISPR/Cas9 technology and the interconnectedness of direct and indirect stakeholders result in a complex decision-making environment for entrepreneurs. James *et al.* (2016) show that entrepreneurs respond to such complexity by drawing on a confirmation or falsification bias. Confirmation bias is an individual's tendency to seek or retain information that is consistent with their favored position and to discount information that is not consistent with their favored position (see also Einhorn and Hogarth, 1978; Schwenk, 1986; Russo and Schoemaker, 1992; Nickerson, 1998; Ng and James, 2016). Such a bias has been observed in entrepreneurial cognition studies that find entrepreneurs tend to develop an egocentric view of their information environment. The result is that they are more likely to rely on information that reaffirms their self-interest or beliefs and to ignore or discount countervailing information than managers in more established businesses (Palich and Bagby, 1995; Baron, 1998; James *et al.*, 2016; Ng and James, 2016). The reason is that entrepreneurs operate in more fluid, uncertain and complex social settings than managers operating in regular firms, without clear lines of decision-making authority and reporting accountability, and thus entrepreneurs are less able to engage in counterfactual reasoning than their managerial counterparts. This bias suggests that entrepreneurs will identify and exploit valuable opportunities by selectively enacting those stakeholder resources and experiences that confirm or validate what they already know. In effect, entrepreneurs are overconfident in their decisions and decision-making ability (Baron, 1998; Hayward *et al.*, 2006). The result is that entrepreneurs tend to favor the interests of direct stakeholders over those of indirect stakeholders, even when information exists that shows harm to indirect stakeholders.

The claim that entrepreneurs favor direct over indirect stakeholders in complex decision-making environments because of a confirmation bias is not the same as saying they favor direct over indirect stakeholders because of raw self-interest. Even if entrepreneurs consciously seek to restrain their self-interests by considering the perspective of indirect stakeholders, the more complex the decision-making environment, the more

persistent the bias will be felt by decision makers. This is a result of cognitive limitations of human decision makers, not their self-interest.

Since the commercialization of the CRISPR/Cas9 technology operates within a complexity of direct and indirect stakeholder exchanges, efforts to commercialize the technology by innovators and entrepreneurial start-up firms may result in key decision makers being vulnerable to confirmation bias. The consequence is that such entrepreneurial efforts may fail to fully account for interests of indirect stakeholders who do not have a direct and immediate stake in the commercial interest. Stated differently, the creation, development and commercialization of technologies place greater demands on entrepreneurial firms to seek the resources, expertise and capital investments of those stakeholders who can bring an entrepreneur's inventions to market. As a result, entrepreneurs will be more likely to focus their attention on the interests of direct stakeholders than on the interests of indirect stakeholders, not only because they have a legitimate stake in bringing its technology to market, but also because of a confirmation bias that results in a greater inclination to give favorable or selective attention to direct stakeholders over indirect stakeholders.

For instance, while the specific targeting and replacement of genes in the CRISPR/Cas9 technology offers the potential to revolutionize medical health and agricultural industries, the ethical consequences of CRISPR/Cas9 to the broader or indirect members of society have yet to be sorted out, but doing so might take a backseat to commercialization efforts. For example, Feng Zhang co-founded Editas Medicine only months after publishing his ground-breaking scientific papers on CRISPR (Byrnes, 2016). In an interview, Zhang said that 'the thing everybody should focus on is how we can push this technology forward, so we can actually treat a disease' (Kolker, 2016, n.p.). Similarly, Ledford quotes another biophysicist as saying that 'there is a mentality that as long as it works, we don't have to understand how or why it works' (2015, p. 21). These statements have serious ethical implications on society.

However, as one commentator observes, 'The company also has to hope that ethical questions about some potential uses of CRISPR do not cloud the technology and delay regulatory approval. For example, some scientists are concerned that the technology could be used to engineer human embryos' (Byrnes, 2016, n.p.). Moreover, recent studies in both mouse subjects and non-viable human embryos have found that CRISPR/Cas9 editing can increase the number of 'off-site mutations' in a cell's DNA. Such off-site mutations can result in unintended and unknown long-term side effects to an organism and its surrounding population. For instance, an RNA-guided 'gene drive' can be devised to eradicate the population of an invasive species, such as mosquitos with dengue fever. Yet, off-target mutations in these mosquitos can increase the potential that their edited genes are transferred to other species (Rodriguez, 2016). This can result in the destruction of other species and thus affect the agricultural food production system. Furthermore, Chinese researchers were recently involved in using the CRISPR/Cas9 editing technique to correct blood genetic disorders (i.e. beta thalassemia) in abnormal human embryos. Their CRISPR/Cas9 system changed some of the targeted genes, while at the same time also created an excessive number of off-site mutations to the embryo's DNA. The research had to be abandoned, with the conclusion that the genome editing technique is not ready for clinical use (Stockton, 2015; Rodriguez, 2016). Simply, unlike somatic cell modification, the use of the CRISPR/Cas9 in the human germline can be passed on from one generation to the next (Stockton, 2015). The result is that unintended off-site target mutations on the human population will be unknown. However, because indirect stakeholders do not have a direct stake in the commercialization of the CRISPR/Cas9 technologies, the impact of the above ethical issues on decisions to commercialize the technology is likely to be subordinate to the commercial interests of the direct stakeholders.

Proposition 1a: The complexity surrounding the commercialization of CRISPR technologies

has a positive influence on an entrepreneur's confirmation bias.

Proposition 1b: Conditional on the complexity of commercializing CRISPR technologies, entrepreneurial start-up firms are vulnerable to an ethical blindness, where the commercial interests of direct stakeholders are given greater priority over the ethical concerns of their indirect stakeholders.

Novelty

Something is novel when the development or application of the thing is new, original or unusual. CRISPR/Cas9 technology is novel not only because it offers an unprecedented ability to target specific genes at a much lower cost than all previous genetic technologies (Booth, 2016), but also because its wide applicability offers various opportunities to combine the technologies and expertise of others into developing and commercializing products that the world has not previously seen.

Entrepreneurs frequently operate in decision settings characterized by novelty (Busenitz and Barney, 1997; Simon and Houghton, 2003; Hayward et al., 2006; McMullen and Shepherd, 2006). Novel decision settings are characterized by a co-creative Schumpeterian process (Schumpeter, 1934) of assimilating and combining the complementary resources and experiences of external partners (Cohen and Levinthal, 1990; Rothaermel and Deeds, 2006; Wiltbank et al., 2006). This novel combination of resources can introduce new commercial applications not previously envisioned by the entrepreneurial firm.

For example, while pesticide- and drought-resistant crops have been achieved through genetic modification, engineering plants that are resistant to disease is a more difficult and challenging matter. This is because a plant's disease-resistance genes cannot be modified significantly, since they can become too active in damaging the plant (Borel, 2017). CRISPR/Cas9 offers the means to target specific portions of the disease-resistant gene without causing changes to other parts of the plant's genome. For example,

CRISPR/Cas9 has been used in creating rice that is resistant to bacterial leaf streak and blight and it can also be developed in ways that alter plants so that they are resistant to a wide range of diseases. For this reason, Monsanto entered into a licensing agreement with scientists at Harvard and MIT universities for the use of a variation of CRISPR technology called CRISPR/Cpf1. According to a Monsanto news release, the company believes the 'system may offer an expanded set of benefits for advancing and delivering improved agricultural products' (Monsanto, 2017, n.p.). Monsanto hopes the technology can be used in developing new generation crops that are disease resistant (Borel, 2017), thus highlighting a novel application of the technology.

Inside bias

A consequence of novel decision-making settings is that decision makers are inclined to adopt an inside bias (Kahneman and Lovallo, 1993; James *et al.*, 2016), which is a tendency to utilize only case-specific information when making forecasts, assessing progress and evaluating impacts. In contrast, an outside view tends to ignore case-specific information and instead favors a consideration of external clues and similar examples. In essence, the difference between inside view and outside view is the source of where decision makers draw their information. According to Kahneman and Lovallo, when 'decision makers are excessively prone to treat problems as unique [they end up] neglecting both the statistics of the past and the multiple opportunities of the future' (1993, p. 17), thus reflecting their susceptibility to novelty-inducing inside bias.

An inside bias has been recognized by scholars in entrepreneurs' judgments of novel product-market decisions (Kahtri and Ng, 2000; Simon and Houghton, 2003; Read *et al.*, 2009). This is because entrepreneurs typically have a personal stake in bringing their innovations to market (Miles *et al.*, 2004) and thus intimately understand the challenges and obstacles of their particular venture's innovation processes. However, because they are so focused on the commercial success of their innovation, they tend to ignore or discount the example of similar ventures and the struggles entrepreneurs faced in those cases.

As novelty can induce an inward bias, we argue that an entrepreneur's efforts to commercialize the CRISPR/Cas9 can yield an 'inside' view of the commercialization process. For example, consider the patent process. A consequence of the inside view is that while the entrepreneurial inventor gains a much-needed understanding of the challenges and processes of the patent process, the entrepreneur's attention becomes biased to those stakeholders who are most affected by their efforts in protecting its patent claims. For instance, studies have found that entrepreneurs tend to be pre-occupied with overcoming the day-to-day challenges of innovation (Simon and Houghton, 2003) and that they are unaware of the potentially harmful impacts their innovation may have to broader members of society (Lepoutre and Heene, 2006; Dew and Sarasvathy, 2007). This suggest that as entrepreneurs are pre-occupied with, say, patent approval, they may gain a deep awareness of how the scope of their patent claims can affect the commercial interest of its direct stakeholder, but they will tend to do so at the expense of its indirect stakeholders. This is because an entrepreneur's efforts to defend and protect their patent claims is dependent on their ability to demonstrate a novel 'practical' application of their technology and not on the welfare and moral concerns of the indirect stakeholder. As a result, when commercializing novel technologies, the ethical consequences surrounding the CRISPR/Cas9 technologies may not be given full consideration by the inventors of this technology. Consequently, novelty can yield an inside bias where the entrepreneurial start-up firm is less likely to feel accountable to their indirect stakeholders.

Proposition 2a: The novelty surrounding the commercialization of CRISPR technologies has a positive influence on an entrepreneur's inward bias.

Proposition 2b: Conditional on the novelty of commercializing CRISPR technologies, entrepreneurial start-up firms are vulnerable to an ethical blindness, where the commercial interests of direct stakeholders are given greater priority over the ethical concerns of their indirect stakeholders.

Ethical Tensions from New Technology Commercialization

The advancement of a society's health and medical wellbeing is often attributed to breakthroughs in medical sciences. While many scientists as well as venture capitalists believe that CRISPR/Cas9 has the potential to dramatically improve the wellbeing of society, such breakthrough innovations also introduce ethical tensions and concerns that may not be fully considered by its inventors and developers. This is because breakthrough innovations introduce a host of commercial issues (i.e. patent disputes, licensing agreements, raising of venture capital, academic credit, prestige) that are of more immediate concern to the valued interest of its inventors than broader ethical concerns raised by indirect stakeholders. Addressing these commercial issues is central to moving CRISPR/Cas9 into the realm of 'practical' use so that society is able to realize the medical and agronomic benefits of this technology.

However, the commercialization of CRISPR/Cas9 exhibits properties of complexity and novelty that can render entrepreneurs vulnerable to confirmation and inward biases. Yet, while cognitive biases can render the entrepreneur vulnerable to ethical blindness with respect to the concerns of indirect stakeholders, they may also be helpful in explaining why some people exploit entrepreneurial opportunities while others do not (Brandstätter, 2011). For example, De Carolis and Saparito state that 'recent research in the field of entrepreneurship suggests several specific cognitive biases that influence risk perception as they relate to entrepreneurs: *overconfidence, illusion of control*, and *representativeness*' (2006, p. 45; emphasis in original). In other words, an important ethical tension might exist between the cognitive and personality traits needed to commercialize CRISPR/Cas9 and those that affect the ability of entrepreneurial innovators to identify with the ethical concerns of indirect stakeholders.

Moreover, while the advancement and commercialization of scientific knowledge are key to bringing new medical and agronomic discoveries to society, this 'technology push' model of science tends also to 'push' out the consideration about the ethical consequences of these advancements. A failure to fully incorporate these ethical considerations into an entrepreneur's commercialization decision can have potentially irreversible long-term and unknown consequences to society. For example, the character Ian Malcolm in the 1993 movie *Jurassic Park* speaks to the issues raised here. In the movie, Ian Malcolm lectures John Hammond, the creator of the dinosaur theme park, about the ethical concerns surrounding the use of genomic technologies in its creation of dinosaurs. He says:

> genetic powers are the most awesome force the world has ever seen, but you yield it like a boy who has found his dad's gun ... I'd tell you the scientific power that is being used here. It didn't require any discipline to attain it. You read what others have done and took the next step, you did not earn the knowledge for yourselves so you didn't take any responsibility for it ... *Your scientists were so preoccupied with whether or not they could, they didn't stop to think if they should.*

Certainly not all scientists developing new technologies and seeking their commercialization, such as those advancing CRISPR/Cas9, are taking an 'anything goes' attitude. For example, Jennifer Doudna expressed concerns surrounding the unregulated use of CRISPR. When she learned about a postdoc's work in engineering a virus to carry CRISPR components into mice that could potentially create a model for lung cancer in humans, she said, 'It seemed incredibly scary that you might have students who were working with such a thing' (quoted in Ledford, 2015, p. 21). Given the relatively low cost of this technique, the potential to misuse this technology in creating, say, lung cancer in humans – for the purpose of weaponization – is not beyond the technical limits of this technology. However, Doudna and other developers of the technology continue to defend the scientific merits of CRISPR/Cas9, even while some in the medical research community have called for a temporary moratorium in its use for germline editing. Doudna made the following comments in regards to a Chinese embryo study:

> I don't think there's anything wrong with their research. The fact that they used triponuclear zygotes, and their work could not in any respect ever become a baby, is an important ethical safeguard. As far as we can tell, they met legal requirements in China. What they did would be illegal in some US states, and they certainly couldn't apply for federal funding. But I don't think they have anything to be ashamed of. I thought it was useful research, but if anything this makes it more real, and makes it look like it may happen more. And even just beyond the fact of how poorly it worked, it did work in a human embryo. This makes it even more urgent to have a societal conversation about how far to go. My own view of the *Science* paper is that it didn't call for a moratorium on all research, it called for a moratorium on making babies this way. (Quoted in Stockton, 2015, n.p.)

Conclusion

Scholars have examined how the decision-making environment can induce an ethical blindness (Palazzo *et al.*, 2012) or bounded awareness (Chugh and Bazerman, 2007) in decision makers. Scholars have also linked psychology and cognition to deteriorating ethical judgments. For example, Tenbrunsel and Messick refer to 'ethical fading' as 'the failure to acknowledge the innate psychological tendency for individuals to engage in self-deception. Individuals do not "see" the moral components of an ethical decision, not so much because they are morally uneducated, but because psychological processes fade the "ethics" from an ethical dilemma' (2004, p. 224). Chugh *et al.* define 'bounded ethicality' as 'limits on the quality of decision making with ethical import' (2005, p. 75). In their view, a 'conflict of interest is not

limited to explicit dishonesty'; rather, context and the cognitive limitations of humans make it difficult for them to recognize and appropriately navigate complicated ethical challenges.

In this chapter we also consider the role of psychology and cognition in identifying and understanding ethical tensions that can arise when developers of new technologies seek to commercialize their innovations. In considering the specific case of the gene editing system known as CRISPR/Cas9, we argue that the existence of complexity and novelty in the decision-making process to commercialize the technology can make entrepreneurial innovators susceptible to confirmation and inside biases that in turn can affect their ability to consider fully the ethical implications on indirect stakeholders. In making this claim we are not suggesting that entrepreneurial innovators will inevitably or always experience deteriorating ethics when they face complex and novel decision-making environments. A confluence of factors affects the ethical sensitivity, judgment and behavior of individuals (see, for instance, Treviño, 1986; Jones, 1991). Moreover, James *et al.* (2016) show that some individuals have characteristics and traits that allow them to operate in complex and novel decision settings in a way that makes them less susceptible to having to deal with cognitive biases. However, it is important to recognize that cognitive biases exist and when and how they arise. If the introduction of new technologies, such as CRISPR/Cas9, is intended to advance the interests of all stakeholders, then decision makers need to be particularly vigilant in ensuring that indirect stakeholders are given appropriate ethical consideration.

References

Baron, R.A. (1998) Cognitive mechanisms in entrepreneurship: Why and when entrepreneurs think differently than other people. *Journal of Business Venturing* 13, 275–294. doi:10.1016/ S0883-9026(97)00031-1

Booth, B. (2016) Riding the gene editing wave: Reflections on CRISPR/Cas9's impressive trajectory. *Forbes*, 31 May. Available at: https://www.forbes.com/sites/brucebooth/2016/05/31/riding-the-gene-editing-wave-reflections-on-crisprs-impressive-trajectory/#7ad343106f9f (accessed 6 November 2017).

Borel, B. (2017) CRISPR, microbes and more are joining the war against crop killers. *Nature*, 14 March. doi:10.1038/543302a

Brandstätter, H. (2011) Personality aspects of entrepreneurship: A look at five meta-analyses. *Personality and Individual Differences* 51, 222–230. doi:10.1016/j.paid.2010.07.007

Busenitz, L.W. and Barney, J.B. (1997) Differences between entrepreneurs and managers in large organizations: Biases and heuristics in strategic decision-making. *Journal of Business Venturing* 12, 9–30. doi:10.1016/S0883-9026(96)00003-1

Byrnes, N. (2016) A big bet that gene editing will cure human disease. *MIT Technology Review*, July 25. Available at: https://www.technologyreview.com/s/601846/a-big-bet-that-gene-editing-will-cure-human-disease/ (accessed 13 November 2017).

Chugh, D. and Bazerman, M.H. (2007) Bounded awareness: What you fail to see can hurt you. *Mind and Society* 6, 1–18. doi:10.1007/s11299-006-0020-4

Chugh, D., Bazerman, M.H. and Banaji, M.R. (2005) Bounded ethicality as a psychological barrier to recognizing conflicts of interest. In: Moore, D.A., Cain, D.M., Loewenstein, G. and Bazerman, M.H. (eds) *Conflicts of Interest: Challenges and Solutions in Business, Law, Medicine, and Public Policy*. Cambridge University Press, Cambridge, pp. 74–95.

Cohen, W.M. and Levinthal, D.A. (1990) Absorptive capacity: A new perspective on learning and innovation. *Administrative Science Quarterly* 35, 128–152. doi:10.2307/2393553

De Carolis, D.M. and Saparito, P. (2006) Social capital, cognition, and entrepreneurial opportunities: A theoretical framework. *Entrepreneurship Theory and Practice* 30, 41–56. doi:10.1111/j.1540-6520.2006.00109.x

Dew, N. and Sarasvathy, S.D. (2007) Innovation, stakeholders and entrepreneurship. *Journal of Business Ethics* 74, 267–283. doi:10.1007/s10551-006-9234-y

DiMasi, J.A., Grabowski, H.G. and Hansen, R.W. (2016) Innovation in the pharmaceutical industry: New estimates of R&D costs. *Journal of Health Economics* 47, 20–33. doi:10.1016/j.jhealeco.2016.01.012

Einhorn, H.J. and Hogarth, R.M. (1978) Confidence in judgment: Persistence of the illusion of validity. *Psychological Review* 85, 395–416. doi:10.1037/0033-295X.85.5.395

Freeman, R.E. (1984) *Strategic Management: A Stakeholder Approach*. Pitman, Boston, Massachusetts.

Hayward, M.L.A., Shepherd, D.A. and Griffin, D. (2006) A hubris theory of entrepreneurship. *Management Science* 52, 160–172. doi:10.1287/mnsc.1050.0483

James, H.S. Jr, Ng, D. and Klein, P. (2016) Complexity, novelty, and ethical judgment by entrepreneurs. *International Journal of Entrepreneurial Venturing* 8, 170–195. doi:10.1504/IJEV.2016.077624

Jones, T.M. (1991) Ethical decision-making in organizations: An issue-contingent model. *Academy of Management Review* 16, 366–395. doi:10.5465/AMR.1991.4278958

Joyner, B.E. and Payne, D.P. (2002) Evolution and implementation: A study of values, business ethics and corporate social responsibility. *Journal of Business Ethics* 41, 297–311. doi:10.1023/A:1021237420663

Kahneman, D. and Lovallo, D. (1993) Timid choices and bold forecasts: A cognitive perspective on risk taking. *Management Science* 39, 17–31. doi:10.1287/mnsc.39.1.17

Kahtri, N. and Ng, H.A. (2000) The role of intuition in strategic decision making. *Human Relations* 53, 57–86.

Klein, P.G. (2008) Opportunity discovery, entrepreneurial action, and economic organization. *Strategic Entrepreneurship Journal* 2, 175–190. doi:10.1002/sej.50

Kolker, R. (2016) How Jennifer Doudna's/Feng Zhang's gene editing technique will change the world. *Bloomberg*. Available at: https://www.bloomberg.com/features/2016-how-crispr-will-change-the-world/ (accessed 7 November 2017).

Ledford, H. (2015) CRISPR, the disruptor. *Nature* 522, 20–24. doi:10.1038/522020a

Lepoutre, J. and Heene, A. (2006) Investigating the impact of firm size on small business social responsibility: A critical review. *Journal of Business Ethics* 67, 257–273. doi:10.1007/s10551-006-9183-5

Lewis, T. (2015) Scientists may soon be able to 'cut and paste' DNA to cure deadly diseases and design perfect babies. *Business Insider*, 19 November. Available at: http://www.businessinsider.com/how-crispr-will-revolutionize-biology-2015-10 (accessed 6 November 2017).

Markman, J. (2017) This is how gene-editing will change the food you eat. *Forbes*, 11 July. Available at: https://www.forbes.com/sites/jonmarkman/2017/07/11/this-is-how-gene-editing-will-change-the-food-you-eat/#35e3ce3c583d (accessed 10 November 2017).

McKinsey & Company (2017) Realizing the potential of CRISPR. Available at: https://www.mckinsey.com/industries/pharmaceuticals-and-medical-products/our-insights/realizing-the-potential-of-crispr (accessed 10 November 2017).

McMullen, J.S. and Shepherd, D.A. (2006) Entrepreneurial action and the role of uncertainty in the theory of the entrepreneur. *Academy of Management Review* 31, 132–152. doi:10.5465/AMR.2006.19379628

Miles, M.P., Munilla, L.S. and Covin, J.G. (2004) Innovation, ethics, and entrepreneurship. *Journal of Business Ethics* 54, 97–101. doi:10.1023/B:BUSI.0000043501.13922.00

Monsanto (2017) Monsanto announces global genome-editing licensing agreement with Broad Institute for newly-characterized CRISPR system. News release, 4 January. Available at: https://monsanto.com/news-releases/monsanto-announces-global-genome-editing-licensing-agreement-with-broad-institute-for-newly-characterized-crispr-system/ (accessed 7 November 2017).

Ng, D. (2011) Thinking outside the box: An absorptive capacity approach to the product development process. *International Food and Agribusiness Management Review* 14, 67–94.

Ng, D. (2015) Entrepreneurial overconfidence and ambiguity aversion: Dealing with the devil you know, than the devil you don't know. *Technology Analysis and Strategic Management* 27, 946–959. doi:10.1080/09537325.2015.1037266

Ng, D. and James, H.S. Jr (2016) No man lives on an island: Habitual agency and complexity in entrepreneurial decision-making. *International Journal of Complexity in Leadership and Management* 3, 244–259. doi:10.1504/IJCLM.2016.080312

Nickerson, R.S. (1998) Confirmation bias: A ubiquitous phenomenon in many guises. *Review of General Psychology* 2, 175–220. doi:10.1037/1089-2680.2.2.175

Noland, J. and Phillips, J. (2010) Stakeholder engagement, discourse ethics and strategic management. *International Journal of Management Reviews* 12, 39–49. doi:10.1111/j.1468-2370.2009.00279.x

Palazzo, G., Krings, F. and Hoffrage, U. (2012) Ethical blindness. *Journal of Business Ethics* 109, 323–338. doi:10.1007/s10551-011-1130-4

Palich, L.E. and Bagby, D.R. (1995) Using cognitive theory to explain entrepreneurial risk taking: challenging conventional wisdom. *Journal of Business Venturing* 10, 425–438. doi:10.1016/0883-9026(95)00082-J

Read, S., Dew, N., Sarasvathy, S., Song, M. and Wiltbank, R. (2009) Marketing under uncertainty: the logic of an effectual approach. *Journal of Marketing* 73, 1–18. doi:10.1509/jmkg.73.3.1

Rodriguez, E. (2016) Ethical issues in genome editing using Crispr/Cas9 system. *Journal of Clinical Research & Bioethics* 7, 266. doi:10.4172/2155-9627.1000266

Rothaermel, F.T. and Deeds, D.L. (2006) Alliance type, alliance experience and alliance management capability in high technology ventures. *Journal of Business Venturing* 21, 429–460. doi:10.1016/j.jbusvent.2005.02.006

Russo, J.E. and Schoemaker, P.J.H. (1992) Managing overconfidence. *Sloan Management Review* 33, 7–17.

Sarasvathy, S. (2001) Causation and effectuation: Toward a theoretical shift from economic inevitability to entrepreneurial contingency. *Academy of Management Review* 26, 243–263. doi:10.5465/AMR.2001.4378020

Schumpeter, J.A. (1934) *The Theory of Economic Development*. Harvard University Press, Cambridge, Massachusetts.

Schwenk, C.R. (1986) Information, cognitive biases, and commitment to a course of action. *Academy of Management Review* 11, 298–310. doi:10.5465/AMR.1986.4283106

Simon, M. and Houghton, S.M. (2003) The relationship between overconfidence and the introduction of risky products: Evidence from a field study. *Academy of Management Journal* 46, 139–149. doi:10.2307/30040610

Stockton, N. (2015) America needs to figure out the ethics of gene editing now. *Wired*, 23 April. Available at: https://www.wired.com/2015/04/america-needs-figure-ethics-gene-editing-now/ (accessed 7 November 2017).

Tenbrunsel, A.E. and Messick, D.M. (2004) Ethical fading: The role of self-deception in unethical behavior. *Social Justice Research* 17, 223–236. doi:10.1023/B:SORE.0000027411.35832.53

Treviño, L.K. (1986) Ethical decision-making in organizations: A person-situation interactionist model. *Academy of Management Review* 11, 601–617. doi:10.5465/AMR.1986.4306235

Wiltbank, R., Dew, N., Read, S. and Sarasvathy, S. (2006) What to do next? The case for non-predictive strategy. *Strategic Management Journal* 27, 981–998. doi:10.1002/smj.555

12 New Technology, Ethical Tensions and the Mediating Role of Translational Research

Corinne Valdivia,[1]* Harvey S. James, Jr.[1] and Roberto Quiroz[2]

[1]*Division of Applied Social Sciences, University of Missouri, Columbia, Missouri, USA; [2]Crop and Systems Sciences Division, International Potato Center, Lima, Peru*

Introduction

Ethical tensions surface as a result of conflicting interests of the different actors, different value systems and rights of various stakeholders, and differences in power and access. All of these often are present simultaneously in smallholder farming communities in developing countries. The interests may appear to be similar between scientists and smallholder farmers because both may care about food security and development. But the way each group thinks about these problems and their solutions can be very different. Both groups make decisions informed by very different experiences, knowledge and positions of power. The values, ideals and the circumstances of smallholder farmers may be securing food and wellbeing in a context of uncertainty and possible risks. Scientists may be aiming to increase productivity or efficiency in the production of a given food crop that can contribute to food security in the context of lab or field experiments where there is no uncertainty and risk of insecurity is measurable and defined. Complicating things further, there may be unintended consequences of scientific solutions because of differences in knowledge, access and power relations between those intended to benefit from the innovation and those who actually do. Indeed, there are often surprises that arise from the complexity of interactions among actors, and the social, economic and environmental contexts. Additionally, the assumptions scientists make in order to solve the problems in developing countries may not play out in the context of smallholder farming systems.

Scientists are seeking new technological innovations that can meet the challenge of food security in the developing world within the context of economic globalization, a changing climate and intensification in more difficult environmental settings. Translational research processes aim to resolve or mediate the potential unintended consequences, ethical tensions and problems that arise from scientific solutions and applications of new technologies, including technological fixes. Translational research processes also provide a framework for addressing transformational changes that create uncertainties and risks that make it difficult to identify new and alternative ways of 'doing'. A key aspect is two-way communication among relevant stakeholders. Elements that characterize the translational research process include: focusing on the context and the relevant stakeholders; identifying partners in the research with the knowledge and understanding of the problems who can facilitate

* Corresponding author. E-mail: valdiviac@missouri.edu

or impede the project and who will be there after the research project is completed; facilitating discovery and learning through participatory research processes that build trust.

The importance of translational research as an approach to resolving or mediating the potential ethical tensions that arise with technological fixes is illustrated in the report on genetically engineered (GE) crops produced by the National Academies of Sciences, Engineering, and Medicine (NAS) in the USA. According to the authors: 'The social and economic effects of GE ... crops depend on the fit of the GE trait and the plant variety to the farm environment and the quality and cost of the GE seeds. GE crops have benefited many farmers on all scales, but genetic engineering alone cannot address the wide variety of complex challenges that face farmers, especially smallholders' (NAS, 2016, p. 2). For this reason, participation of relevant stakeholders is essential. If ethical tensions arise when interests, rights and values are in conflict, then including the participation of important stakeholders in the development and implementation of new technology can be an effective means of mitigating such tensions.

Some approaches have been developed to address ethical tensions from new technology introduced in developing countries, such as the development of communities of practice in South Africa (Hendrickson *et al.*, 2014). Others, finding that farmers' voices are not being heard, propose the need for different research methodologies and outreach (Wedding and Tuttle, 2013). Indeed, in a continent where the vast majority of producers are small-scale farmers, the voices of the farmers themselves remain largely muted within this debate. Those who have worked for years connecting with farmers in developing countries point to the benefits of participatory research (Hayward *et al.*, 2004). Participatory research allows farmers and scientists to develop a common set of expectations and vocabulary to discuss alternative strategies. By participating in research, farmers can make their own observations and can derive lessons from research beyond those conclusions presented by the researcher. The

translational research approach has been applied to the development of adaptation practices in the context of climate change in the Andes (Valdivia *et al.*, 2010), tensions about biotechnology in East Africa (Valdivia *et al.*, 2014) and more recently to the development of remote-sensing technologies for smallholder cropping systems in Africa using a platform of unmanned aerial vehicles or drones (CIP, 2016). The approach has addressed many ethical tensions between different stakeholder groups that belong to the practice of farming, engaged in a process that facilitates or enhances the voice of smallholder farmers.

This chapter presents two cases that relate to innovations in smallholder contexts, where food insecurity and a changing climate create a context of vulnerability framed by uncertainty and risks. The innovations – genetically modified cassava for smallholder farming in Kenya and remote-sensing technologies with unmanned aerial vehicles for East African smallholder farmers – provide examples of how translational research can address some of the ethical issues that arise from technological fixes. The common thread in these cases is the way participatory approach can help mitigate potential ethical challenges that arise when scientists and innovators seek solutions to difficult problems in contexts where there are differences in values, power and knowledge by a diversity of actors or stakeholders and institutions, and where the context itself may impact on the outcomes.

Case 1: Genetically Modified Crops and Farmer Voice in the Context of Smallholder Farming

Technological innovations are needed to meet the challenge of food security in a developing world where populations will continue to grow and rapid environmental and climate changes are occurring. A statement made in 1999 by US Secretary of Agriculture Dan Glickman about biotechnology is informative: 'with all that technology has to offer, it is nothing if it's not accepted. This boils down to a matter of trust ... in the science behind the process, but particularly trust in the regulatory process that ensures thorough review – including complete and open public involvement' (quoted in NAS, 2016, p. x). Smallholder farmers will adopt new technology if they understand and trust the science and scientists involved in its development. But for farmers to do this they need to be active rather than passive participants in the development and implementation of new technology, and this means they need to have a voice. According to James and Sulemana:

> For farmers to have a voice, they must be empowered and active participants, not passive recipients. They must be equal to other stakeholders participating in decisions about objectives, methods, scope and dissemination plans. By farmers we mean primarily smallholder farmers, particularly those in developing countries, although lessons here could apply to other marginalized groups. (2014, p. 638)

An initiative funded by the Templeton Foundation focused on biotechnology and smallholder farmers in the developing world (see Mitton and Bennett, 2015). Several of these projects provide insight into the importance and relevance of farmer voice. In Uganda, Schnurr and Mujabi-Mujuzi (2014) found that scientists often make simplifying assumptions about farmers and their motives and preferences, such as only seeking to maximize yields and profits. Farmers also typically learn about the agricultural biotechnology only when it is completely developed. In Mexico, Carro-Ripalda and Astier (2014) observed that the GM maize controversy has resulted in a silencing of the voices of farmers against GM, pointing to structural issues of power that should not be ignored. In contrast, in South Africa, Hendrickson *et al.* (2014) found that because GM maize is available and dominates commercial production, there was an opportunity to establish a Community of Practice with black South African smallholders entering farming to learn about hybrid and GM maize production. The Community of Practice that engaged with stakeholders and farmers lifted their voices in some areas but not others,

such as with issues involving storage and transportation. They also found that bringing stakeholders together in non-hierarchical ways encouraged new ways of thinking and new partnerships that help scientists better understand the concerns and needs of farmers, moving away from the simplifying assumptions about what farmers want.

A more detailed examination of GM cassava in Kenya provides additional insight into the importance of smallholder farmer voice in the development of and conversation about GM crops and how translational and participatory research processes can facilitate this process.

Cassava plays an important role as a food security crop for smallholder farmers in sub-Saharan Africa. As a food, some cassava varieties can be eaten fresh; others are boiled, or processed into chips or flour. Cassava can also be processed to remove the starch (which in turn can be processed into tapioca and other food products) and used as a biofuel and animal feed. It is an effective food security crop, because it is drought tolerant, it is a low-input crop and it can be stored in the ground for long periods of time. However, two diseases threaten cassava: cassava mosaic disease (CMD) and cassava brown streak disease (CBSD). These diseases can result in yield losses of 80% (Walsh *et al.*, 2012). The Donald Danforth Plant Sciences Center in St Louis, Missouri, in cooperation with the Kenya Agricultural Research Institute (KARI, but today the Kenyan Agricultural and Livestock Research Organization), partnered in the development of a genetically modified cassava that is resistant to CMD and CBSD (Taylor *et al.*, 2012).

Given this context, two authors (Valdivia and James) established collaborations with KARI in 2013 in order to examine the benefits, risks and unintended consequences of introducing GM cassava in Kenya. The translational research process we utilized included all stakeholders currently engaged in some way with cassava, including producers, processors, researchers, extension officers, other organizations working in research and development, retailers and farmers in different regions of Kenya, in order to capture the diversity of cultures, roles of cassava in food and livelihoods, access to extension and other stakeholders, as well as degrees of food insecurity and vulnerability. KARI helped us select sites for farmer focus groups in the Coast and Eastern Provinces where cassava is an important crop for food security or there is potential for commercialization, in rural villages where people are food insecure, and where our collaborators in KARI have research centers nearby. Two sites were selected in each region to capture cultural and agroecological differences: Kwale and Kilifi for the Coast region and Machakos and Makueni for the Eastern region. Two of the four sites were well connected to organizations providing information about new technologies, including agricultural biotechnology, whereas the other two were less connected. A multiple embedded case study design was used to learn about the different practices, uses and roles of cassava from the perspectives of men and women. The research process consisted of participatory research with farmer groups to learn about their vulnerability context, the role of cassava for food security and livelihoods, their concerns about their livelihoods, as well as their preferences about cassava varieties, and their knowledge or awareness of GM crops. Researchers engaged extension officers in the fieldwork. A second stage consisted of sharing with cassava scientists at KARI the knowledge, preferences and concerns of farmers. A third step was for scientists to participate in workshops with farmers' groups to further learn and share and to answer concerns farmers raised in the participatory research. At these workshops KARI scientists also organized sessions to present information that responded to issues identified by farmers regarding cassava varieties, processing and marketing, which are major issues from the farmers' perspectives. To complement this research, stakeholders (such as religious leaders, processors, regulators and scientists) were interviewed to learn their perspectives about GM cassava. Finally, farmer representatives elected by each group attended a final workshop in Nairobi with officials from each region, researchers and other stakeholders, which included farmer associations, government officials, the media and

members of the existing network on GMOs, to share findings and discuss other relevant issues.

The general conclusion reached from the study is that there would not be significant barriers to the adoption by farmers in Kenya of GM crops generally and GM cassava specifically. The main concern smallholder farmers have is ensuring that the right varieties of cassava are genetically modified. Farmers are particular about the varieties of cassava they use. GM cassava that is clean (i.e. free from mosaic or brown streak disease) but that is not the type farmers traditionally use or need will not be accepted by them. This is one reason why farmer voice is essential. Scientists working on genetically modifying crops need to know the varieties of crops farmers and consumers want and will use.

Another conclusion reached from the study is that smallholder farmers will follow the advice of opinion makers. Key opinion makers for farmers are KARI (now KALRO) and agricultural extension workers. Other opinion leaders are farmer groups, local community leaders and religious leaders, as well as government leaders. It is important for opinion makers to have an accurate understanding of biotechnology and to be able to communicate risks and benefits accurately to farmers. How information is presented to farmers is important. Thus, communication is the key issue here. From the perspective of farmers, their voice is heard through an iterative process with scientists responding to their concerns, such as the traits needed in crops as well as issues relating to marketing, inputs needed with new varieties and other matters relating to smallholder farming, and then reporting back to them what they did or can do with respect to genetic modification.

Importantly, potential unintended consequences of new technology implementation can be identified through translational and participatory research. An example related to gender from our work in Kenya is that women market their cassava products in small amounts, while men tend to market larger quantities of production. If the focus on genetic modification is simply increasing yields, then it may have a gender effect by changing the spaces for cassava production and processing. Another challenge for women is harvesting larger roots. The networks women participate in are not useful for accessing information about varieties or better markets for these kinds of products.

This case illustrates the importance of translational and participatory research methods in the development and implementation of new technology, especially agricultural biotechnology. Furthermore, scientists learned that GM adoption is feasible and even desirable under the right conditions, but knowing what those conditions are requires the voice of those who are adopting the GM crops, such as smallholder farmers. The translational process facilitates the exchange between farmers and scientists.

Why is it important for smallholder farmers to have a voice? Farmers will ultimately decide which crops to grow, and how to take care of their food security needs, as they are in charge of negotiating their unique contexts. Participatory approaches can contribute to building knowledge, changing perceptions, identifying barriers and creating coalitions among stakeholders. Another benefit of this type of collaboration is that it strengthens the human capital and expands the networks of collaboration in an organization whose mandate is 'to promote, streamline, coordinate and regulate all aspects of research in agriculture and livestock development, and also promote the application of the research findings and technologies in the country' (KALRO, 2018, n.p.).

Case 2: Unmanned Aerial Vehicles in East Africa

In many tropical regions, scientists, agriculturalists and farmers want to understand the characteristics of various landscapes for research and planning, but issues with cloud cover during the rainy and growing seasons pose a particular challenge. In these situations, it is important to obtain accurate information of topographical areas, crop acreage, health of crops and yield forecasting in order to improve the quality of information about farming conditions and food crops and to facilitate timely and cost-effective decision making. In developing countries, there are significant food security implications involved here.

One method of facilitating this is through the use of unmanned aerial vehicles (UAVs) or drones. A project led by another author (Quiroz) at the International Potato Center (in cooperation with author Valdivia) sought to develop a low-cost UAV-based agricultural remote sensing information system (ARSIS) to survey crops in sampling areas of smallholder farming in East Africa (see CIP, 2016). Multispectral cameras mounted on UAV platforms register reflected radiation from vegetation in the green, red and near infrared regions of the electromagnetic spectrum in gridded pixels of less than 5 cm per side. At this spatial resolution, crops can be discriminated from each other through digital processing of the images and accurate crop statistics can be collated. The purpose of the project was to determine if this technology could generate information that would ultimately benefit smallholder farmers. This 'proof of concept' project used a translational research approach to engage with developers of the technology, applications by scientists and understanding by potential users.

Conceptually, the translational approach involved a path from knowledge to action and was based on three concepts: (i) Community of Practice (CoP); (ii) participatory processes to engage with the knowledge systems of a diversity of stakeholders; and (iii) outcomes and impacts process (learning) measured through participatory workshops. The purpose was to identify how this innovation could become a tool used in the production of information by decision makers whose practice seeks to improve smallholder farming and food security. Two-way communications processes were used to facilitate mutual learning about UAV-ARSIS.

A CoP is an ideal method to identify the needs of stakeholders and build a network to address challenges or barriers, which in this case means involving organizations that focus on the development of smallholder farming in Africa. A CoP consists of 'groups of people informally bound together by shared expertise and passion for a joint enterprise' (Wenger and Snyder, 2000, p. 139). An important part of the CoP is the sharing of knowledge and expertise through regular interactions, thus the primary 'output' of the CoP is knowledge. However, there are other benefits of CoPs that have application to ethical tensions relating to the introduction of new technology because the identification and sharing of common interests is a key aspect of the way a CoP operates. Stated differently, the CoP is one where multiple stakeholders are engaged in shaping how the innovation becomes a relevant tool for practice, facilitating or contributing research on the ground, informing their own organizations as trusted sources of information, and identifying how in each context the innovation can move forward, exploring its potential in each field and developing the capacities to sustain it.

Figure 12.1 represents the UAV-ARSIS CoP and the key objectives of learning, building trust and networks, and identifying impact pathways. The CoP comprises those who develop the technology, those who can apply the innovation in their own research and who can use the products in their own

practice, those who facilitate and enable its access and use, and users of the information derived from its use (see also Valdivia and Quiroz, 2014). This is relevant in Africa where history, cultures, policies and institutions are very diverse. Therefore, understanding the perspectives of stakeholders is essential in ensuring that the innovation is salient to them, that it is actionable and trusted, and that it is developed in the appropriate institutional context (rules of the game) of each African country. Engaging with application scientists – in this case the national agricultural research institutions and extension program operations, conducting trials and working with farmers – are key elements in developing and testing the innovation. The feedback loops depict a dynamic process. They also show that in each context where the innovation process takes place there are different stakeholders and rules. The CoP is essential because the network consists of a diversity of stakeholders, each in a different context (organizations and countries), with a shared understanding of UAV-ARSIS, who can communicate and work together to identify pathways that enable it. Stakeholders participate at every stage of the process, from the generation

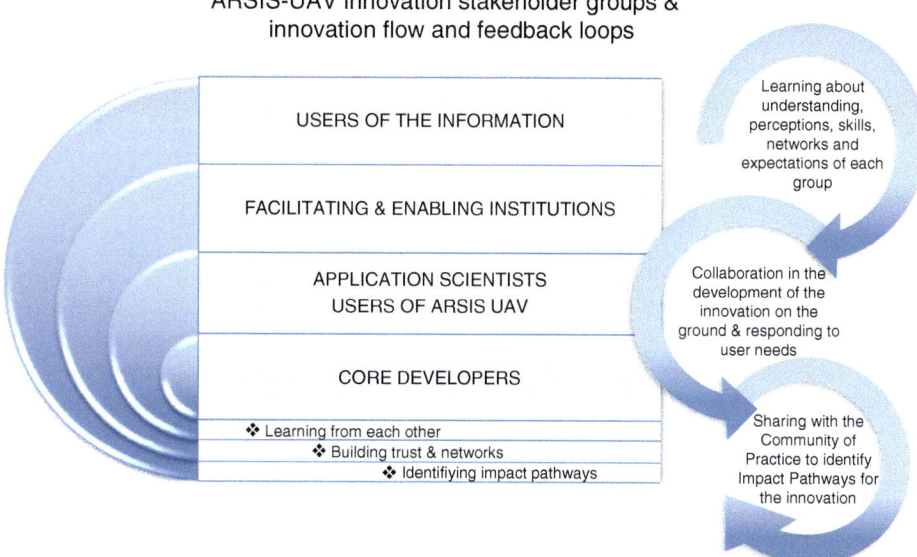

ARSIS-UAV innovation stakeholder groups & innovation flow and feedback loops

USERS OF THE INFORMATION

FACILITATING & ENABLING INSTITUTIONS

APPLICATION SCIENTISTS
USERS OF ARSIS UAV

CORE DEVELOPERS

❖ Learning from each other
❖ Building trust & networks
❖ Identifiying impact pathways

Learning about understanding, perceptions, skills, networks and expectations of each group

Collaboration in the development of the innovation on the ground & responding to user needs

Sharing with the Community of Practice to identify Impact Pathways for the innovation

Fig. 12.1. Illustration of the UAV-ARSIS Community of Practice. (Valdivia and Quiroz, 2014.)

and adaptation of the innovation, to testing it in the field, enabling its application and the development of information for decision makers.

The CoP was the organic entity through which the translational research process occurred. Translational research focuses on processes that facilitate effective communication of knowledge, wants and needs, as well as concerns, and informs on the innovation itself. This is particularly important when innovations may be perceived differently. For example, UAVs (drones) in Africa can be a concern to potential adopters due to the fact that these are identified with military operations and because terrorist attacks in Nairobi generated panic in the governments of the region. In fact, as private-sector businesses were beginning to engage in the use of drones (Zirulnick, 2015), the Kenyan government issued a ban on UAVs due to the government's concern about a terrorist group. This presented a challenge to the field-testing in the region. The Ugandan government had similar concerns at the time the research collaborators were getting ready to test UAV-ARSIS with sweet potato crops. Hence, lack of trust and perceived risks were a challenge to overcome. The concerns over drone use illustrate an important ethical tension that new technology can create when introduced in these contexts. If smallholder farmers and other users or observers of technology attribute it to things they fear or do not understand, then negative perceptions will impede adoption. These need to be addressed directly before anticipated benefits can be understood and accepted. Communication and building trust, which are hallmarks of translational research processes, are essential for overcoming them.

The communication elements of translational research process can develop meaningful and actionable knowledge. Meaningful knowledge is a knowledge that is relevant, in the language and context of the decision maker, and delivered by a trusted source. While these are necessary conditions for the movement of knowledge to action, they often are not sufficient. In order to be actionable, networks and institutions that facilitate its use are needed. To develop meaningful

knowledge there are at least three elements to take into account. The first is that two-way participatory communication can enhance trust (Wilkins, 2001). The second is the use of workshops using participatory approaches, as these have been found to be effective in facilitating communication in Africa (Patt *et al.*, 2005). The third is that context matters, which means that it is important to work together with local decision makers, as they understand the context and contribute their perspectives on how the innovation can be effective (Valdivia *et al.*, 2014).

For this reason, participatory workshops were implemented to engage stakeholders in the innovation and use of UAV-ARSIS. An inception workshop in October 2014 in Nairobi, Kenya, engaged stakeholders from Tanzania, Uganda, Kenya, other East African organizations (National Agricultural Research Organization and farmer associations) and six international agriculture research centers with mandates in Africa, Asia and Latin America, including the International Potato Center (CIP). At this workshop, developers presented the state of the art in civil use of drone-based remote sensing and their views about what UAV-ARSIS could mean for each stakeholder group's practice. The workshop also facilitated the development of networks of collaboration for the necessary fieldwork testing and identified issues raised by stakeholders needing answers. The importance of the inception workshop was in bringing together key stakeholders, to learn about their interests, concerns, knowledge and perceptions, and the relevance of UAV-ARSIS and how it could contribute to their practice. Take-away lessons included the necessity of involving local institutions at different stages, testing the technology in generating information or making official statistics, employing a stepwise process from simple to complex tools, complementarity with satellite imageries and recognizing the importance of the regulatory process.

Developers listened to the feedback from different stakeholders and redefined project priorities to respond to the perceived needs. An outcome was the establishment of collaborations with local partners in the development and testing of UAV-ARSIS on the ground. Development of UAV-ARSIS was strengthened by collaborations with universities and government agricultural institutions in Kenya, Tanzania, Uganda and Rwanda, which also facilitated field-testing and learning about the challenges in enabling the innovation in the region. Importantly, this also meant that engagement in human capital would remain in each country well after the life of the project, with people trained in open access technologies and low-cost services, in order to continue efforts to make the technology economically viable and thus available for both smallholder farmers as well as large commercial operations.

The second workshop in June 2016 in Nairobi, Kenya, convened stakeholders who had participated in the first workshop, and others working with remote-sensing technologies and UAVs in Mali and Tanzania, as well as private-sector organizations from Nigeria and the USA, and a European institution (the Technical Centre for Agricultural and Rural Cooperation). The central question addressed by stakeholders in this workshop was whether UAVs could 'become part of the tools that improve agriculture and the wellbeing of smallholder farmers in Africa'. The translational process entailed learning from the stakeholders, responding to their questions, and developing actionable knowledge and information that addressed their needs. The workshop informed on the progress achieved and responded to issues raised in the first workshop. It also included a technology fair that was organized for hands-on learning on UAV technology and for sharing communication products developed in response to identified needs. Training videos were developed in collaboration with the University of Missouri journalism program, as well as videos for the public that documented fieldwork collaborations (O'Brien 2015, 2016). Blogs and articles were prepared and published (Allen, 2015), and an online CoP was facilitated (UAV4Ag; see CTA, no date).

Another objective was to learn from experiences about the innovation and approval process in various countries in Africa. An important finding from this part of the study was that a majority of countries do not have a regulatory framework in place to

manage issues relating to the use of UAVs in agriculture (see, for instance, Jeanneret and Rambaldi, 2016). Some countries have regulations that differentiate use for research vs commercial purposes. Other countries have banned the use of UAVs because of security concerns. Many pointed out the need to engage Civil Aviation Regional Authorities, which while invited did not attend either of the workshops. There is a need to raise awareness specifically about the use of drones in agriculture, and what UAV-ARSIS is about, since users and policy makers might perceive UAVs in agriculture as drones that harm crops or people or crops. This is a key area that needs to be addressed. For example, we found that as a result of the interactions within the CoP and workshops, developers noted that communicating with farmers addressed the potential of blaming the technology (UAVs) for issues with crops that may be related to diseases or pests; these are concerns that were not in the minds of the developers initially. We also found that engaging potential users in a CoP facilitated testing in Tanzania and Rwanda. In other words, it was the CoP that ultimately made possible field missions to showcase the utility of UAV-ARSIS, because the relationships established at the first workshop developed interest and trust that made possible approvals for field-testing UAVs in both countries. Representatives from these countries, as well as Uganda and Kenya, shared their experiences, raising the need to focus on communications to inform those working to develop regulations.

As further evidence of the ability of CoPs and translational processes to affect change, questions on rating knowledge before and after the workshop were used to assess change in knowledge. The second workshop contributed to changing knowledge about UAV-ARSIS in 82% of participants. Of these, almost 50% changed their knowledge from poor or average to excellent. Almost 50% of participants changed from none or poor knowledge to average; and 50% changed one level, while 32% changed by two levels. Thus, the potential for work such as this to address or alleviate concerns and ethical tensions from new technology, such as UAVs, can be validated.

Participants were also asked to identify next steps they would like to see happen after the second workshop. Developers identified the need to communicate and find ways to change the mentality of stakeholders about the UAVs, build capacity through partnerships between universities and research centers, and develop a project that encompasses several crops and organizations so that benefits realized from the overall project could expand and grow. Application scientists identified the need to learn more about specifics of the innovation through collaboration with developers, to assess costs and benefits from continued collaboration across stakeholder groups and prospects for building capacities in the technology, and to consolidate a CoP to collaborate and contribute communication and advocacy. Users of the products and those in enabling environments identified the need to continue stakeholder forums, collaborate in experimental trials with UAV-ARSIS, and work as a community to advocate for the innovation. This stakeholder group also identified the need for capacity-building related to operation and commercial access of UAVs, and to participate in planning and designing future projects and programs aimed at upscaling the innovation across eastern Africa. The needs identified for the different stakeholder groups are consistent with their 'practices'. Their perspectives contribute to developing the steps in the process of UAV-ARSIS becoming an actionable information technology for smallholder agriculture.

Overall, the translational research process utilized in the development of UAVs for agricultural use in Africa helped to overcome important ethical tensions that the new technology created, or could have created. The feedback process informed on how the technology would be useful to potential users and how to communicate issues and concerns about UAV-ARSIS with different stakeholders. It also addressed the risks and potential fears of UAVs among public officers as well as users and thus facilitated the field-testing of UAV-ARSIS. Finally, the translational process created a network of stakeholders that continues beyond the life of the project. This has the added benefit of providing a mechanism for addressing new ethical tensions or challenges that might arise later as the technology continues to be developed and adopted.

Conclusion

New technologies can bring about ethical tensions due to the differences in interests, values, rights, power and access of stakeholders, especially in the development of smallholder agriculture. Our two cases illustrate how translational research processes can mitigate ethical tensions. A key aspect of translational research is the facilitation of two-way communication among relevant stakeholders, through communities of practice and other means of getting stakeholders together and promoting the sharing of ideas, interests and knowledge.

While both scientists and smallholder farmers care about food security and development, their solutions can be very different. For instance, as illustrated in the case of GM cassava in Kenya, scientists thought to increase yields by reducing disease exposure, but smallholder farmers had other concerns relating to their livelihoods and food security. Through the translational research process, these differences became evident early on, because farmer voice was made salient to scientists developing the technology. Furthermore, as illustrated in the case of unmanned aerial vehicles in East Africa, the fears of adopters about harmful uses of technologies or misperceptions about what technology can and cannot do will limit its appeal. Translational research processes, especially through communities of practice, helped provide necessary knowledge and understanding to adopters as well as policy makers and enhanced awareness of concerns by technology developers. This allowed important challenges to testing and adoption to be overcome.

Thus, we conclude that while translational research will not solve all ethical tensions that new technology creates, it can be effective in mitigating important challenges that arise from conflicting, or misunderstood, interests and concerns of stakeholders.

References

Allen, W. (2015) CIP drone study over sweet potato fields of East Africa a success. Available at: http://cipotato.org/press-room/blog/cip-drone-study-over-sweetpotato-fields-of-east-africa-a-success/ (accessed 28 February 2018).

Carro-Ripalda, S. and Astier, M. (2014) Silenced voices, vital arguments: Smallholder farmers in the Mexican GM maize controversy. *Agriculture and Human Values* 31, 655–663. doi:10.1007/s10460-014-9533-3

CIP (2016) *Remote Sensing as a Monitoring Tool for Smallholder's Cropping Area Determination in Tanzania and Uganda using Sweet Potato as a Pilot Crop*. Final Report submitted to The Bill and Melinda Gates Foundation. CIP, Lima.

CTA (n.d.) Unmanned Aerial Vehicles for Agriculture (UAV4Ag) Community of Practice workshop. CTA, Wageningen, Netherlands. Available at: http://www.cta.int/en/article/2016-06-06/unmanned-aerial-vehicles-for-agriculture-uav4ag-community-of-practice-workshop.html (accessed 28 February 2018).

Hayward, C., Simpson, L. and Wood, L. (2004) Still left out in the cold: Problematising participatory research and development. *Sociologia Ruralis* 44, 95–108. doi:10.1111/j.1467-9523.2004.00264.x

Hendrickson, M.K., Gilles, J.L., Meyers, W.H., Schneeberger, K.C. and Folk, W.R. (2014) Choice and voice: Creating a community of practice in KwaZulu-Natal, South Africa. *Agriculture and Human Values* 31, 665–672. doi:10.1007/s10460-014-9532-4

James, H.S. Jr and Sulemana, I. (2014) Case studies on smallholder farmer voice: An introduction to a special symposium. *Agriculture and Human Values* 31, 637–641. doi:10.1007/s10460-014-9554-y

Jeanneret, C. and Rambaldi, G. (2016) Drone governance: A scan of policies, laws and regulations governing the use of unmanned aerial vehicles (UAVs) in 79 ACP countries. CTA Working Paper Series ICTs for Agriculture No. 16/12. October. Available at: http://publications.cta.int/en/publications/publication/1971/ (accessed 28 February 2018).

KALRO (2018) Kenya Agricultural & Livestock Research Organization: Vision and Mission. Available at: http://www.kalro.org/vision-and-mission (accessed 28 February 2018).

Mitton, P. and Bennett, D. (eds) (2015) *Analysis: Africa's Future … Can Biosciences Contribute?* Banson/B4FA, Cambridge. Available at: http://b4fa.org/wp-content/uploads/2016/02/B4FA-Analyses-complete.pdf (accessed 6 July 2018).

National Academies of Sciences, Engineering, and Medicine (NAS) (2016) *Genetically Engineered Crops: Experiences and Prospects*. NAS, Washington, DC. doi:10.17226/23395

O'Brien, C. (2015) Airborne agriculture. An international team of scientists is developing low-cost, open-source drone technology to monitor crops. 22 December. Available at: http://wildtech.mongabay.com/2015/12/airborne-agriculture/ (accessed 28 February 2018).

O'Brien, C. (2016) In the sky, under the earth: Drones and sweet potatoes in Tanzania. Presentation at Workshop II UAV ARSIS, Nairobi Kenya, 7 June. Available at: https://youtu.be/qzZCCohN_4Y (accessed 28 February 2018).

Patt, A., Suarez, P. and Gwata, C. (2005) Effects of seasonal climate forecasts and participatory workshops among subsistence farmers in Zimbabwe. *Proceedings of the National Academy of Sciences of the United States of America* 102, 12623–12628. doi:10.1073/pnas.0506125102

Schnurr, M.A. and Mujabi-Mujuzi, S. (2014) 'No one asks for a meal they've never eaten.' Or, do African farmers want genetically modified crops? *Agriculture and Human Values* 31, 643–648. doi:10.1007/s10460-014-9537-z

Taylor, N.J., *et al.* (2012) The VIRCA project: Virus resistant cassava for Africa. *GM Crops and Food: Biotechnology in Agriculture and the Food Chain* 3: 93–103. doi:10.4161/gmcr.19144

Valdivia, C. and Quiroz, R. (2014) UAV ARSIS Inception Workshop 14 October. International Livestock Research Institute, Nairobi, Kenya. Workshop Report. CIP, Lima Peru.

Valdivia, C., *et al.* (2010) Adapting to climate change in Andean ecosystems: Landscapes, capitals, and perceptions shaping rural livelihood strategies and linking knowledge systems. *Annals of the Association of American Geographers* 100, 818–834. doi:10.1080/00045608.2010.500198

Valdivia, C., *et al.* (2014) Using translational research to enhance farmers' voice: A case study of the potential introduction of GM cassava in Kenya's coast. *Agriculture and Human Values* 31, 673–681. doi:10.1007/s10460-014-9536-0

Walsh, S., Odero, B.O. and Obiero, H. (2012) C3P on-farm vouchers: Pilot use of on-farm vouchers to disseminate cassava planting material in Western Kenya. C3P Crop Crisis Control Project, Brief 5. Available at: https://www.crs.org/sites/default/files/tools-research/c3p-pilot-use-of-on-farm-vouchers-to-disseminate-cassava-planting-material.pdf (accessed 28 February 2018).

Wedding, K. and Tuttle, J.N. (2013) *Pathways to Productivity. The Role of GMOs for Food Security in Kenya, Tanzania, and Uganda*. A Report of the Center for Strategies and International Studies

CSIS Global Food Security Project. Bowman & Littlefield, Lanham, Maryland.

Wenger, E.C. and Snyder, W.M. (2000) Communities of practice: The organizational frontier. *Harvard Business Review* (Jan–Feb), 139–145.

Wilkins, L. (2001) A primer on risk: An interdisciplinary approach to thinking about public understand of agbiotech. *AgBioForum* 4, 163–172.

Available at: http://agbioforum.org/v4n34/v4n34a03-wilkins.htm (accessed 5 July 2018).

Zirulnick, A. (2015) Kenya was set to be a perfect lab for commercial drones until regulators struck. *Quartz Africa*, 15 May. Available at: https://qz.com/413370/kenya-was-set-to-be-a-perfect-lab-for-commercial-drones-until-regulators-struck/ (accessed 28 February 2018).

Index

CABI – who we are and what we do

This book is published by **CABI**, an international not-for-profit organisation that improves people's lives worldwide by providing information and applying scientific expertise to solve problems in agriculture and the environment.

CABI is also a global publisher producing key scientific publications, including world renowned databases, as well as compendia, books, ebooks and full text electronic resources. We publish content in a wide range of subject areas including: agriculture and crop science / animal and veterinary sciences / ecology and conservation / environmental science / horticulture and plant sciences / human health, food science and nutrition / international development / leisure and tourism.

The profits from CABI's publishing activities enable us to work with farming communities around the world, supporting them as they battle with poor soil, invasive species and pests and diseases, to improve their livelihoods and help provide food for an ever growing population.

CABI is an international intergovernmental organisation, and we gratefully acknowledge the core financial support from our member countries (and lead agencies) including:

Discover more

To read more about CABI's work, please visit: **www.cabi.org**

Browse our books at: **www.cabi.org/bookshop**,
or explore our online products at: **www.cabi.org/publishing-products**

Interested in writing for CABI? Find our author guidelines here:
www.cabi.org/publishing-products/information-for-authors/